Lecture Notes in Mathematics 1517

Klaus Keimel Walter Roth

Ordered Cones and Approximation

Springer-Verlag
Berlin Heidelberg New York
London Paris Tokyo
Hong Kong Barcelona
Budapest

Authors

Klaus Keimel
Fachbereich Mathematik
Technische Hochschule Darmstadt
Schloßgartenstr. 7
W-6100 Darmstadt, Germany

Walter Roth
Department of Mathematics
University of Bahrain
P. O. Box 32038
Isa Town
State of Bahrain

Mathematics Subject Classification (1991): 41-02, 41A36, 41A65, 46A22, 47H04

ISBN 3-540-55445-9 Springer-Verlag Berlin Heidelberg New York
ISBN 0-387-55445-9 Springer-Verlag New York Berlin Heidelberg

Typesetting: Camera ready by author/editor
Printing and binding: Druckhaus Beltz, Hemsbach/Bergstr.
46/3140-543210 - Printed on acid-free paper

Table of Contents

Introduction

Korovkin type approximation theorems typically deal with certain restricted classes of continuous linear operators on locally convex vector spaces. These may be positive operators on ordered vector spaces or contractions on normed spaces as in the seminal work by Korovkin [29], [30], Shashkin [50], [51], Wulbert [63], Bauer [9], and Behrens and Lorentz [12]. Less known situations include operators on Banach algebras [4], [39], [40] or on spaces of stochastic processes [60], [61] with certain restricting properties. More recently, there are results about order preserving linear operators on set-valued functions [59], [28], [13], [14] as well. In this case the domain of the operators under consideration is no longer a vector space but only a cone, i.e. subtraction of elements is not always defined (see also [44], [45]). Generally, if an approximation process is modelled by such a restricted class of operators, those very restrictions guarantee convergence towards the identity on a large subset of their domain if this property may be checked for a relatively small test set. Korovkin's classical theorem (see [29], [30]) states that a sequence of positive linear operators on $C[0,1]$ converges towards the identity for all functions in $C[0,1]$ if this holds for the three "test functions" $1, x$ and x^2. Unfortunately, the different situations mentioned above and the different restrictions on the classes of operators so far required different approaches and techniques.

Looking for a unified presentation of Korovkin type approximation theorems we had to leave the setting of vector spaces and turn to more general structures which we call locally convex cones. For our purposes, it is essential to include cones which are not embeddable in vector spaces. As we need to apply functional analytic concepts, in particular an appropriate duality theory and Hahn-Banach type extension and separation theorems, we have to stay reasonably close to the classical theory of locally convex vector spaces, yet allow sufficient generality in order to serve our main purpose: Various restrictions on classes of operators in Korovkin type approximation may be taken care of by the proper choice of domains and their topologies alone. Thus, we just have to investigate continuous operators between locally convex cones. The following is an outline of the main concepts and some (simplified) results of this work.

Preordered cones. A *cone* is a set P endowed with an addition $(a,b) \to a+b$ and a scalar multiplication $(\alpha, a) \to \alpha a$ for real numbers $\alpha \geq 0$. The addition is only supposed to be associative and commutative, and a neutral element 0 is required to exist (see Ch.I.1).

Of course, cones in real vector spaces are cones in the above sense. They have the *cancellation property*

$$a+c = b+c \ \text{ implies } \ a = b$$

which we do not require in general.

In addition we shall assume that P carries a *preorder*, i.e. a reflexive transitive relation "$\leq$" such that

$$a \leq b \text{ implies } a+c \leq b+c \text{ and } \alpha a \leq \alpha b \text{ for all } a,b,c \in P \text{ and all } \alpha \geq 0.$$

Locally convex cone topologies. Our model for introducing a locally convex topology on P is the set $Conv(E)$ of all non-empty convex subsets of a locally convex topological vector space E. It has a natural addition and a scalar multiplication by non-negative reals; it is ordered by inclusion. An arbitrary base $V \subset Conv(E)$ of convex neighborhoods of 0 in E induces three hyperspace topologies on $Conv(E)$ given by the respective neighborhood bases for $A \in Conv(E)$:

in the *upper topology* $\quad V(A) = \{B \in Conv(E) \mid B \subset A+V\},\ V \in V,$

in the *lower topology* $\quad (A)V = \{B \in Conv(E) \mid A \subset B+V\},\ V \in V,$

in the *symmetric topology* $\quad V(A) \cap A(V),\quad V \in V.$

Identifying elements of E with singleton sets, E is naturally embedded in $Conv(E)$, and all the three topologies on $Conv(E)$ coincide with the given locally convex topology on E.

For an abstract formulation we use order theoretical concepts to introduce locally convex topologies on cones (see I.2.2): A subset V of the cone P is called an ***(abstract) 0-neighborhood system***, if the following properties hold:

$0 < v$ for all $v \in V$;

for all $u,v \in V$ there is $w \in V$ with $w \leq u$ and $w \leq v$;

$u+v \in V$ and $\alpha v \in V$ whenever $u,v \in V$ and $\alpha > 0$.

For every $a \in P$ we define

$$v(a) = \{b \in P \mid b \leq a+v\}$$

to be a neighborhood of a in the *upper topology*, and

$$(a)v = \{b \in P \mid a \leq b+v\}$$

to be a neighborhood of a in the *lower topology*. The common refinement of these two topologies is called the *symmetric topology* on P. We call (P,V) a *full locally convex cone.* We also consider subcones Q of P not necessarily containing V. They will be endowed with the topologies induced from Q and denoted as *locally convex cones* (Q,V). They are the general subject of our study.

Cones are asymmetric structures, so asymmetric conditions come as no surprise: For technical reasons we require the elements of a locally convex cone to be *bounded below,* i.e. for every $a \in Q$ and $v \in V$ we have $0 \leq a+\rho v$ for some $\rho > 0$.

Note that the upper neighborhoods $v(a)$ are decreasing convex sets and the $(a)v$ are increasing convex sets. The neighborhoods in the symmetric topology are both convex and order convex. Thus, all of these three topologies merit to be called locally convex. Of course, the upper and the lower topology are far from being Hausdorff. Since all of the three topologies are defined in terms of the preorder on Q, continuity properties etc. will be expressible by means of the ordering and the 0-neighborhoods alone.

The global preorder (see I.3.9). On a locally convex cone (Q,V) we define the ***global preorder*** "$\preceq$" for $a,b \in Q$ by

$$a \preceq b \quad \text{if and only if} \quad a \leq b+v \quad \text{for all} \quad v \in V.$$

It is easy to check that $a \leq b$ for our original preorder always implies $a \preceq b$. Furthermore, mappings between locally convex cones which are continuous either with respect to their upper or their lower topologies need to be monotone with respect to their global preorders. This is the main reason for the usefulness of locally convex cones in Korovkin type approximation theory.

The following standard examples will be used throughout our text:

Examples. (a) Clearly every locally convex topological vector space E with 0-neighborhood base V is a locally convex cone (E,V) in this sense. (E is a subcone of the full locally convex cone $(Conv(E),V)$.)

(b) The cone $\overline{R} = R \cup \{+\infty\}$ may be endowed with the abstract neighborhood system $V = \{\varepsilon \in R \mid \varepsilon > 0\}$. For $a \in R$ the intervals $(-\infty, a+\varepsilon)$ are the upper and the intervals $(a-\varepsilon, +\infty)$ the lower neighborhoods, while for $a = +\infty$ the entire cone $\overline{R}$ is the only upper neighborhood, and $\{+\infty\}$ is open in the lower topology. The symmetric topology on $\overline{R}$ is the usual topology on R with $\{+\infty\}$ as an isolated point.

(c) Let $(E,\leq)$ be a locally convex ordered topological vector space with 0-neighborhood base V. For $a,b \in E$, and $V \in V$ we define

$$a \leq b+V \quad \text{if there is some} \quad v \in V \quad \text{such that} \quad a \leq b+v.$$

Thus (E,V) is a locally convex cone and the symmetric topology on E coincides with the original one if the neighborhoods $V \in V$ are order convex.

(d) If (Q,V) is a locally convex cone then there is a canonical way to define a locally convex topology on the cone $Conv(Q)$ of non-empty convex subsets of Q:
For convex sets $A,B \in Conv(Q)$ and a neighborhood $v \in V$ we set

$$A \leq B+\overline{v} \quad \text{if and only if for all} \quad a \in A \quad \text{there is some} \quad b \in B \quad \text{such that} \quad a \leq b+v.$$

Thus $(Conv(Q),\overline{V})$ is a locally convex cone (we set $\overline{V} = \{\overline{v} \mid v \in V\}$). It does not satisfy the cancellation property.

Moreover, it may be shown that every locally convex cone satisfying a minor additional assumption (see Theorem II.2.20) admits a representation as a subcone of $(Conv(E),\overline{V})$, where (E,V) denotes a suitable locally convex topological vector space.

(e) If (Q,V) is a locally convex cone and X is a compact space by $C(X,Q)$ we denote the cone of Q-valued functions on X which are continuous with respect to the symmetric topology on Q. For $f,g \in C(X,Q)$ and $v \in V$ we set

$$f \leq g+\overline{v} \quad \text{if and only if} \quad f(x) \leq g(x)+v \quad \text{for all} \quad x \in X.$$

Thus, endowed with this topology of uniform convergence $(C(X,Q),\overline{V})$ is a locally convex cone as well.

(f) In the context of Korovkin type approximation theorems for linear contractions on normed spaces the following locally convex cone is of interest: Let $(E,\|\ \|)$ be a normed vector space with unit ball B. Let $Q = \{a+\rho B \mid a \in E,\ \rho \geq 0\}$ be provided with the canonical neighborhood basis $V = \{\rho B \mid \rho > 0\}$ and the set inclusion as preorder.

Uniformly continuous linear operators (see II.1.1 and II.1.2). For cones Q and P, a map $T : Q \to P$ is called a *linear operator*, if

$$T(a+b) = T(a)+T(b) \quad \text{for all } a,b \in Q \text{ and}$$
$$T(\alpha a) = \alpha T(a) \quad \text{for all } a \in Q \text{ and } \alpha \geq 0.$$

If (Q,V) and (P,W) are locally convex cones then the linear operator $T : Q \to P$ is called *uniformly continuous* or *u-continuous* for short, if for every $w \in W$ one can find a $v \in V$ such that

$$a \leq b+v \quad \text{implies} \quad T(a) \leq T(b)+w.$$

Uniform continuity is not just continuity. It is immediate from the definition that it implies and combines continuity with respect to the upper, lower and symmetric topologies on Q and P. Every u-continuous linear operator is monotone with respect to the global preorder. In Example (f), for example, if we extend a given linear operator T on the normed space E to a linear operator $\overline{T}$ on $Q = \{a+\rho B \mid a \in E,\ \rho \geq 0\}$ by setting $\overline{T}(B) = B$, then u-continuity is equivalent for T to be contractive.

The dual cone (see II.2). By the above a linear functional $\mu : Q \to \overline{R}$ is u-continuous if there is a neighborhood $v \in V$ such that

$$a \leq b+v \quad \text{implies} \quad \mu(a) \leq \mu(b)+1.$$

The u-continuous linear functionals on Q form again a cone, denoted by Q^* and called the *dual cone* of Q. We endow Q^* with the topology $w(Q^*,Q)$ of pointwise convergence of the elements of Q, considered as functions on Q^* with values in $\overline{R}$ with its usual topology. The *polar* v_Q° of a neighborhood $v \in V$ consists of all linear functionals fulfilling the above condition. It is seen (II.2.4) to be $w(Q^*,Q)$-compact and convex.

The following is derived using Hahn-Banach type theorems as in [22]:

Extension Theorem (II.2.9). *Let P be a subcone of the locally convex cone (Q,V). Then every u-continuous linear functional on P can be extended to a u-continuous linear functional on Q; more precisely: For every $\mu \in v_P^\circ$ there is a $\tilde{\mu} \in v_Q^\circ$ such that $\mu = \tilde{\mu}|_P$.*

Superharmonicity with respect to a subcone. Introducing the notation for an element of a locally convex cone to be superharmonic with respect to a given subcone, in Chapter III we turn to applications in approximation theory. This notion is well-known and useful in classical Korovkin theory. It serves the same purpose in our more general setting:

Let Q_0 be a subcone of the locally convex cone (Q,V). Let $\mu \in Q^*$ and $a \in Q$. We shall say that the element a is *Q_0-superharmonic in* μ (see III.1.1) if firstly $\mu(a)$ is finite and if secondly, for all $\nu \in Q^*$,

$$\nu(b) \leq \mu(b) \text{ for all } b \in Q_0 \text{ implies } \nu(a) \leq \mu(a).$$

In order to give an example how classical results may be transferred to our more general concept, recall the following well-known statement in the $C_0(X)$-case which is due to Bauer and Donner [11]:

Theorem. *Let X be a locally compact space and G a linear subspace of $C_0(X)$, the space of continuous real-valued functions on X vanishing at infinity. For a function $f \in C_0(X)$ the following conditions are equivalent:*

(i) For every net $(T_\alpha)_{\alpha \in A}$ of equicontinuous positive linear operators on $C_0(X)$,

$$T_\alpha(g) \to g \text{ for all } g \in G \text{ implies } T_\alpha(f) \to f.$$

(ii) For every $x \in X$ we have

$$f(x) = \sup_{\varepsilon > 0} \inf \{g(x) \mid g \in G, f \leq g + \varepsilon\} = \inf_{\varepsilon > 0} \sup \{g(x) \mid g \in G, g \leq f + \varepsilon\}.$$

(iii) For every $x \in X$ and every bounded positive Radon measure μ on X,

$$\mu(g) = g(x) \text{ for all } g \in G \text{ implies } \mu(f) = f(x).$$

Convergence in (i) is meant with respect to the topology of uniform convergence on $C_0(X)$.

Consider the locally convex cone $(C_0(X),V)$ where V consists of the strictly positive constant functions on X. The dual cone then is formed by the bounded positive Radon measures on X. Clearly, condition (iii) in the preceding theorem means that both f and $-f$ are G-superharmonic in all point evaluations of X. Thus, the equivalence of (ii) and (iii) is a consequence of our

Sup-Inf-Theorem (III.1.3). *Let Q_0 be a subcone of the locally convex cone (Q,V). Let $a \in Q$ and $\mu \in Q^*$ such that $\mu(a)$ is finite. Then a is Q_0-superharmonic in μ if and only if*

$$\mu(a) = \sup_{v \in V} \inf\{\mu(b) \mid b \in Q_0,\ a \leq b + v\}.$$

Like its classical counterpart this theorem may be used to derive Stone-Weierstraß type theorems for locally convex cones (see III.3.6 and III.3.7).

In Chapter IV we turn to Korovkin type approximation theory in locally convex cones and introduce techniques involving adjoint operators (see II.2.15) on the dual cones. We use the above notation of superharmonicity in order to derive our General Convergence Theorem IV.1.13. For a net $(a_\alpha)_{\alpha \in A}$ in a locally convex cone (Q,V) and an element $a \in Q$ we shall denote by $a_\alpha \uparrow a$ the convergence of (a_α) towards a with respect to the upper topology (IV.1.7). We state a simplified version of this theorem which is however sufficient to derive the classical results including the above by Bauer and Donner (c.f. [4], [11], [19], [20], [49], [61], etc.):

Convergence Theorem (IV.1.11). *Let Q_0 be a subcone of the locally convex cone (Q,V). Suppose that for all $v \in V$ the element $a \in Q$ is Q_0-superharmonic in all elements of the $w(Q^*,Q)$-closure of the extreme points of v_Q°. Then for every net $(T_\alpha)_{\alpha \in A}$ of equicontinuous linear operators on Q*

$$T_\alpha(b) \uparrow b \quad \text{for all } b \in Q_0 \text{ implies } T_\alpha(a) \uparrow a.$$

Our text is not meant to be a complete survey of Korovkin type approximation theory. We do not repeat the classical cases in detail. They have been dealt with in many places and we globally refer to Donner's book [19] on the subject and to the very useful bibliographies and summaries by Altomare/Campiti [7] and Pannenberg [41]. We shall, however, in Chapter IV.2 give a few examples that may indicate how to apply our general results to various classical situations. Chapters V and VI are devoted to further applications of our General Convergence Theorem, most of which are new. We consider cone-valued functions which generate locally convex cones in various ways. In particular, we generalize the concept of Nachbin spaces (see V.1) which is described in [43] and which formalizes the concept of weighted approximation. We thus extend Korovkin type theorems to problems of weighted approximation far beyond those by Gadzhiev in [23], [24]. To give an example of our new results we formulate a Korovkin type theorem for set-valued functions (V.3.2 and V.3.4). For finite dimensional vector spaces, $X = [0,1]$ and $M = \{1, x, x^2\}$ it is due to Vitale [59]. A more general version for the finite dimensional case is contained in [28]:

For a locally convex vector space E with neighborhood base V we denote by $CConv(E)$ the locally convex cone of all non empty compact convex subsets of E (c.f. Example (d) from above). Finally, for a compact space X, $C(X,CConv(E))$ denotes the locally convex cone of all continuous (with respect to the symmetric topology) $CConv(E)$-valued functions. It is ordered by (pointwise) inclusion. We consider convergence with respect to the symmetric topology on $C(X,CConv(E))$ (c.f. 4(e)); i.e. the net $(f_\alpha)_{\alpha \in A}$ of functions in $C(X,CConv(E))$ converges to $f \in C(X,CConv(E))$ if and only if for all $V \in V$ there is some $\alpha_0 \in A$ such that

$$f_\alpha(x) \subset f(x)+V \text{ and } f(x) \subset f_\alpha(x)+V \text{ for all } x \in X \text{ and } \alpha \geq \alpha_0.$$

Recall that a *Korovkin system for* $C(X)$ is a subset M of $C(X)$ such that, for the linear subspace G spanned by M, every function in $C(X)$ fulfills the equivalent conditions of the Theorem quoted on page 5.

In a similar way we say that a subset $\overline{M}$ of $C(X,CConv(E))$ is a *Korovkin system for* $C(X,CConv(E))$ if for every $f \in C(X,CConv(E))$ and every net $(T_\alpha)_{\alpha \in A}$ of equicontinuous monotone linear operators on $C(X,CConv(E))$

$$T_\alpha(g) \to g \quad \text{for all } g \in \overline{M}, \quad \text{implies } T_\alpha(f) \to f.$$

Theorem. *Let E be a locally convex vector space, X a compact space and M a Korovkin system for $C(X)$ consisting of positive functions. Let $\mathcal{U}$ be a subset of $CConv(E)$ such that $0 \in U$ for all $U \in \mathcal{U}$ and $\cup\{\lambda U \mid U \in \mathcal{U}, \lambda \geq 0\} = E$. Then the set-valued functions*

$$x \to g(x)U \ : \ X \to \mathcal{Q}, \quad g \in M, \ U \in \mathcal{U}$$

together with the constant functions

$$x \to C \ : \ X \to \mathcal{Q}, \quad C \in CConv(E)$$

form a Korovkin system for $C(X, CConv(E))$.

Chapter VI investigates quantitative approximation theory using the full strength of our General Convergence Theorem. We derive results on the order of convergence for Korovkin type approximation processes on cone-valued functions. These situations include real- and set-valued functions (see VI.4.3, VI.4.5 and VI.4.6) and stochastic processes (VI.4.10). Most of our results are either new or considerable generalizations of known ones.

Chapter I: Locally Convex Cones

In the first three sections of this chapter we present the basic definitions of ordered and locally convex cones and their associated topologies. In Section 4 we discuss the question of embeddability in locally convex vector spaces. Roughly speaking, we show that the subcone B_Q of *bounded elements* (see 2.3) of a locally convex cone Q is embeddable in a canonical way into a locally convex ordered vector space. In particular, if a locally convex cone happens to be a vector space, then, endowed with its *symmetric topology* (see 2.2), it is a locally convex vector space. Thus, our notion of local convexity reduces to the usual notion of local convexity in the case of vector spaces.

In Section 5 we present an alternative approach for locally convex cones through quasiuniform structures. Indeed, we believe that this approach is the most appropriate and natural one; its disadvantage is that (quasi)uniform structures are less appealing to our intuition. Quasiuniform structures had been introduced under the name of "semiuniform structures" by L. Nachbin [34] as a common generalization of order and uniform structures. In vector spaces one has a canonical uniform structure induced by any vector space topology, and we believe that this uniformity is essential for all the basic facts of functional analysis. Thus, for ordered cones, the appropriate locally convex structure should be defined in terms of quasiuniform structures which carry the information both about order and topology.

1. Cones and preordered cones.

1.1 Cones. We define a *cone* to be a set P endowed with an addition $(a,b) \to a+b$ and a scalar multiplication $(\alpha, a) \to \alpha a$ for real numbers $\alpha > 0$. The addition is only supposed to be associative and commutative and a neutral element 0_P (shortly 0) is required to exist, i.e.:

$$\begin{aligned} (a+b)+c &= a+(b+c) && \text{for all } a,b,c \in P, \\ a+b &= b+a && \text{for all } a,b \in P, \\ 0+a &= a && \text{for all } a \in P. \end{aligned}$$

For the scalar multiplication we require as usual:

$$\begin{aligned} \alpha(\beta a) &= (\alpha\beta)a && \text{for all } \alpha,\beta > 0 \text{ and } a \in P, \\ (\alpha+\beta)a &= \alpha a+\beta a && \text{for all } \alpha,\beta > 0 \text{ and } a \in P, \\ \alpha(a+b) &= \alpha a+\alpha b && \text{for all } \alpha > 0 \text{ and } a,b \in P, \\ 1 \cdot a &= a && \text{for all } a \in P. \end{aligned}$$

In this definition of a cone P, the scalar multiplication is only required to be defined for real numbers $\alpha > 0$. We may - and we shall do this in the sequel - extend the scalar multiplication to $\alpha = 0$ by defining $0 \cdot a = 0$ for all $a \in P$, and all of the above rules remain valid. On the other hand, $\alpha \cdot 0 = 0$ for all $\alpha > 0$ is a consequence of these rules. Indeed, for all $a \in P$ we have

$$a = \alpha(\alpha^{-1}a + 0) = a + \alpha 0,$$

whence $\alpha \cdot 0 = 0$ by the unicity of the neutral element.

1.2 Subcones and faces. A subset Q of a cone P is called a *subcone* if

$$a+b \in Q \quad \text{and} \quad \alpha a \in Q \qquad \text{for all } a,b \in Q \text{ and } \alpha \geq 0.$$

Note that every subcone of P contains 0. A *face* F is a subset of P such that

$$a+b \in F \quad \text{implies} \quad a,b \in F \qquad \text{for all } a,b \in P.$$

Of course, cones in real vector spaces are cones in the above sense. They have the *cancellation property*

(C) $\qquad a+c = b+c \quad$ implies $\quad a = b$

for arbitrary elements a,b,c. Conversely, cones which satisfy the cancellation property are embeddable in real vector spaces. It is important to note that cones in our sense are in general far from being embeddable in vector spaces, as the addition is not supposed to be cancellative. This is essential, as we want to include examples like the following:

1.3 Example. With its straightforward addition and multiplication with $\alpha \geq 0$, the set $\overline{R} = R \cup \{+\infty\}$ is a cone.

1.4 Example: Cones of convex sets. Let P be a cone. A subset A of P is called *convex*, if $\alpha a+(1-\alpha)b \in A$, whenever $a,b \in A$ and $0 \leq \alpha \leq 1$.

We denote by $Conv(P)$ the set of all non-empty convex subsets of P. With the addition and scalar multiplication defined as usual by

$$A+B = \{a+b \mid a \in A \text{ and } b \in B\} \quad \text{for } A,B \in Conv(P),$$

$$\alpha A = \{\alpha a \mid a \in A\} \quad \text{for } A \in Conv(P) \text{ and } \alpha \geq 0,$$

it is easily verified that $Conv(P)$ is again a cone. Convexity is required to show that $(\alpha+\beta)A$ equals $\alpha A+\beta A$: Clearly $(\alpha+\beta)A$ is a subset of $\alpha A+\beta A$. To show the converse, consider an arbitrary element $c \in \alpha A+\beta A$; it can be written $c = \alpha a+\beta b$ with $a, b \in A$; as

$$c = (\alpha+\beta)\left(\frac{\alpha}{\alpha+\beta}a + \frac{\beta}{\alpha+\beta}b\right); \qquad \text{(the case } \alpha = \beta = 0 \text{ is trivial.)}$$

and as

$$\frac{\alpha}{\alpha+\beta}a + \frac{\beta}{\alpha+\beta}b \in A \quad \text{by the convexity of } A,$$

we conclude that $c \in (\alpha+\beta)A$.

Note that every subcone Q of P is convex and satisfies $Q+Q=Q$. In particular, the non-empty convex subsets of a real vector space form a cone in our sense which is far from being cancellative.

1.5 Example: Cones of cone-valued functions. Let P be a cone, X any a set. For P-valued functions on X the addition and scalar multiplication may be defined pointwise. The set $F(X,P)$ of all such functions then is a cone in our sense. But again, the addition is in general not cancellative, as it is not in P.

1.6 Preordered cones. A *preordered cone* is a cone P with a reflexive transitive relation $\leq$ such that

$$a \leq b \text{ implies } a+c \leq b+c \text{ and } \alpha a \leq \alpha b \text{ for all } a,b,c \in P \text{ and all } \alpha \geq 0.$$

If $\leq$ is in addition antisymmetric, i.e. $\leq$ is a partial ordering, then P is called an *ordered cone.*

Examples of preordered cones are $\overline{R}$ with its usual order (Ex.1.3), the set $Conv(P)$ of non-empty convex subsets of a cone P ordered by inclusion (Ex 1.4) and if P is a preordered cone, the set of P-valued functions on a given set X endowed with the pointwise ordering (Ex. 1.5). Every cone P is preordered by its *natural preorder* defined by $a \leq_n b$ if $a+c=b$ for some $c \in P$.

Convex sets in cones may look rather peculiar. For example in $\overline{R}$ all the two element sets $\{a,+\infty\}$ are convex. This phenomenon is somehow remedied by considering only increasing or decreasing sets, or more generally convex sets that are also order convex:

1.7 Example: Cones of decreasing convex sets. A subset a of a preordered cone is called *decreasing*, if $a \in A$ and $b \leq a$ for some $b \in P$ imply $b \in A$. For a subset B of P we denote by:

$$\downarrow B = \{a \in P \mid a \leq b \text{ for some } b \in B\},$$

the decreasing subset generated by B. In a dual way one defines the notion of an *increasing* subset and $\uparrow B$, the increasing subset generated by B. It is easily verified, that $\downarrow B$ and $\uparrow B$ both are convex, whenever B is convex. We denote by $DConv(P)$ the set of all non-empty decreasing convex subsets of P.

For a decreasing convex set A and $\alpha > 0$, the set αA is also decreasing and convex. But $A+B$ need not be decreasing, if A and B are. We therefore modify the addition on $DConv(P)$ and define

$$A \oplus B = \downarrow(A+B) = \{c \in P \mid c \leq a+b \text{ for some } a \in A,\ b \in B\}.$$

With this addition and the usual scalar multiplication $DConv(P)$ becomes a cone ordered by inclusion; the set $\{0\}$ acts as the additive zero element. There is a natural map

$$a \to \downarrow\{a\} \text{ of } P \text{ into } DConv(P),$$

which is order preserving. It is an embedding, i.e. injective, if and only if the preorder on P is in fact an order. If not, the image

$$\downarrow P = \{ \downarrow\{a\} \mid a \in P \}$$

is called the *ordered cone associated with* P.

2. Locally convex cones.

We want to endow our cones with a locally convex structure. Our definition will be guided by the example of the cone $Conv(E)$ of all non-empty convex subsets of a locally convex topological vector space E. In this case we can choose a base $\mathcal{V}$ of convex neighborhoods V of 0 in E. We may suppose that $U+V \in \mathcal{V}$ and $\alpha V \in \mathcal{V}$ whenever $U,V \in \mathcal{V}$ and $\alpha > 0$. Note that $\mathcal{V}$ is a "cone without zero" down directed towards 0. A neighborhood base for a convex set A is given by the sets

$$A + V, \quad V \in \mathcal{V}$$

This induces three hyperspace topologies on $Conv(E)$ given by the respective neighborhood bases for $A \in Conv(E)$

in the *upper topology* $V(A) = \{B \in Conv(E) \mid B \subset A+V\}, \; V \in \mathcal{V}$,
in the *lower topology* $(A)V = \{B \in Conv(E) \mid A \subset B+V\}, \; V \in \mathcal{V}$,
in the *symmetric topology* $V(A) \cap A(V), \quad V \in \mathcal{V}$.

On a subcone Q of $Conv(E)$, we shall consider the induced topologies, also in the case where Q does not contain $\mathcal{V}$. On $Q = E$ all the three topologies coincide with the given locally convex topology.
We now present an abstract formulation:

2.1 Abstract 0-neighborhood systems. Bases. Let P be a preordered cone. A subset $\mathcal{V}$ of P is called an *(abstract) 0-neighborhood system*, if the following properties hold:

$0 < v$ for all $v \in \mathcal{V}$;

for all $u,v \in \mathcal{V}$ there is $w \in \mathcal{V}$ with $w \le u$ and $w \le v$;

$u+v \in \mathcal{V}$ and $\alpha v \in \mathcal{V}$ whenever $u,v \in \mathcal{V}$ and $\alpha > 0$.

One could say that $\mathcal{V}$ is a "subcone without zero directed towards 0". The elements

$$a+v, \quad v \in \mathcal{V},$$

may be called *abstract neighborhoods* of $a \in P$.

A subset $\mathcal{U}$ of an (abstract) 0-neighborhood system $\mathcal{V}$ is called a *base of* $\mathcal{V}$ if for every $v \in \mathcal{V}$ there is a $u \in \mathcal{U}$ and an $\alpha > 0$ such that $\alpha u \le v$.
For a subset $\mathcal{U}$ of Q to be the base for some (abstract) 0-neighborhood system it is necessary and sufficient that it satisfies the following conditions:

$0 < u$ for all $u \in \mathcal{U}$;

for all $u,v \in \mathcal{U}$ there is $w \in \mathcal{U}$ and an $\alpha > 0$ with $\alpha w \le u$ and $\alpha w \le v$.

Such a set U will be called an *(abstract) 0-neighborhood base*. The set of all finite sums $\sum_{i=1}^{N} \alpha_i u_i$ with $\alpha_i > 0$ and $u_i \in U$ is a 0-neighborhood system V and U is a base thereof.

2.2 Locally convex topologies. Let P be a cone with a *0*-neighborhood system V. For every $a \in P$ we define

$$v(a) = \{b \in P \mid b \le a+v\}$$

to be a neighborhood of a in the *upper topology*, and

$$(a)v = \{b \in P \mid a \le b+v\}$$

to be a neighborhood of a in the *lower topology*. One easily verifies that these neighborhood systems define indeed topologies on P. The common refinement of these two topologies is called the *symmetric topology* on P.

Note that the $v(a)$ are decreasing convex sets and then $(a)v$ are increasing convex sets. The neighborhoods in the symmetric topology are both convex and order convex. Thus, all of these three topologies merit to be called locally convex. Of course the upper and the lower topologies are far from being Hausdorff. Since all of the three topologies are defined in terms of the preorder on P, we will not need to work with them explicitly. Continuity properties etc. will be expressible by means of the ordering and the *0*-neighborhoods alone.

We also consider subcones Q of P not necessarily containing V. They will be endowed with the topologies induced from P. Thus, for $a \in Q$, the neighborhood bases for the upper and lower topologies on Q will be given by

$$v_Q(a) = v(a) \cap Q = \{b \in Q \mid b \le a+v\}, \quad v \in V$$

$$(a)_Q v = (a)v \cap Q = \{b \in Q \mid a \le b+v\}, \quad v \in V,$$

respectively.

2.3 Bounded elements. Let P be a preordered cone with a *0*-neighborhood system V. For $v \in V$, an element $a \in P$ is called

upper v-bounded, if $a \le \alpha v$ for some $\alpha > 0$,

lower v-bounded, if $0 \le a + \beta v$ for some $\beta > 0$,

v-bounded, if it is both lower and upper v-bounded.

An element a is called *upper (lower) bounded*, if it is upper (lower) bounded for every $v \in V$. The following is straightforward:

2.4 Proposition. *The set $B(v)$ of all upper v-bounded and likewise the set B of all upper bounded elements is a decreasing subcone in P.*

Now we are ready for our main definition which proceeds in two steps:

2.5 Locally convex cones. Let P be a preordered cone and $V \subset P$ an abstract *0*-neighborhood system. The pair (P,V) is called a *full locally convex cone*, if every element of P is lower bounded. In a full locally convex cone the bounded elements therefore coincide with the upper bounded ones.

Finally, a *locally convex cone* is a pair (Q,V), where Q is a subcone and V the 0-neighborhood system of some full locally convex cone (P,V). Of course, Q inherits the preorder and the topological structure of P. We have neglected to indicate P in the notation, as only those elements of P play a role for Q which are of the form $a+v$ with $a \in Q$ and $v \in V \cup \{0\}$, and these elements form a subcone of P already containing V. But one has to keep in mind that Q in general does not contain the 0-neighborhood system.

Clearly every locally convex topological vector space E with 0-neighborhood base V is a locally convex cone (E,V) in this sense, as it is a subcone of the full locally convex cone $Conv(E)$ ordered by inclusion, which contains V and in which every element is bounded below. The preorder induced on E is just the equality.

If, on the other hand, the locally convex cone Q is a vector space, i.e. contains all negatives of its elements, then $b \leq a+v$, if and only if $b-a \leq v$. All elements of Q are bounded, as they and their negatives are bounded below, and the neighborhoods of 0 with respect to the symmetric topology

$$\begin{aligned} v_0 = (0)v \cap v(0) &= \{b \in Q \mid b \leq v \text{ and } 0 \leq b+v\} \\ &= \{b \in Q \mid b \leq v \text{ and } -b \leq v\} \end{aligned}$$

form the basis for a locally convex vector space topology on Q (not necessarily Hausdorff). But it is obvious that distinct abstract neighborhood systems V in our sense may lead to the same symmetric topology. So even in the case of vector spaces there is a substantially larger variety of locally convex cone topologies than vector space topologies.

The following are our standard examples for locally convex cones to which we will frequently refer in the sequel:

2.6 Example. The cone $\overline{R} = R \cup \{+\infty\}$ will always be endowed with the abstract neighborhood system $V = \{\varepsilon \in R \mid \varepsilon > 0\}$. For $a \in R$ the intervals $(-\infty, a+\varepsilon)$ are the upper and the intervals $(a-\varepsilon, +\infty)$ the lower neighborhoods, while for $a = +\infty$ the entire cone $\overline{R}$ is the only upper neighborhood, and $\{+\infty\}$ is open in the lower topology. The symmetric topology on $\overline{R}$ is the usual topology on R with $\{+\infty\}$ as an isolated point.

2.7 Example. Let $(E,\leq)$ be a locally convex ordered topological vector space with 0-neighborhood base V. For $A,B \in Conv(E)$, the cone of non-empty convex subsets of E, we define

$$A \leq B \text{ if for every } a \in A \text{ there is some } b \in B \text{ such that } a \leq b.$$

Since $Conv(E)$ contains V, and all its elements are bounded below in this sense, $(Conv(E),V)$ is a full locally convex cone. E may be considered as a subcone of $Conv(E)$, hence (E,V) is a locally convex cone. Note that the upper neighborhoods of $a \in E$ contain all elements smaller then a, the lower neighborhoods all larger ones. The symmetric topology on E coincides with the original one if the neighborhoods $v \in V$ are order convex.

2.8 Example. Let (P,V) be a full locally convex cone. If we identify the elements of V with singleton sets $\bar{v} = \{v\}$, then $\bar{V} = \{\bar{v}\}$ is a subset of $Conv(P)$, which can be preordered using the preorder of P. For $A,B \in Conv(P)$ we define

$$A \leq B \text{ if for every } a \in A \text{ there is some } b \in B \text{ such that } a \leq b.$$

Since its elements clearly are bounded below as are the elements of P, $(Conv(P),\bar{V})$ becomes a full locally convex cone.

If (Q,V) is a locally convex cone, i.e. a subcone of some full locally convex cone (P,V), then $Conv(Q)$ is a subcone of $Conv(P)$, hence for any cone D of non-empty convex subsets of Q, the pair $(D,\bar{V})$ is a locally convex cone.

We shall also consider the families $DConv(Q)$ and $\overline{DConv(Q)}$ of decreasing convex subsets, respectively closed decreasing convex subsets of Q, where closure is meant with respect to the lower topology on Q. If we slightly modify the addition (c.f. Example 1.7), both sets will become cones as well:

$$A \oplus B = \downarrow(A+B) \qquad \text{for } A,B \in DConv(Q),$$
$$A \bar{\oplus} B = \overline{\{\downarrow(A+B)\}} \qquad \text{for } A,B \in \overline{DConv(Q)},$$

where $\overline{\{\downarrow(A+B)\}}$ denotes the closure with respect to the lower topology. (We shall see in the following section that the closure in this topology is decreasing for any subset of Q, so the downward arrow in the definition of $\bar{\oplus}$ may in fact be omitted.) With the preorder and the abstract neighborhood system induced by $Conv(Q)$ (for decreasing sets this preorder coincides with inclusion), both $(DConv(Q),\bar{V})$ and $(\overline{DConv(Q)},\bar{V})$ then are locally convex cones.

2.9 Example. Let $F(X,P)$ be the cone of P-valued functions on the set X, where (P,V) is a full locally convex cone. If we consider its pointwise preorder and identify the elements $v \in V$ with the constant functions $x \to v$ for all $x \in X$, then V is a subset of $F(X,P)$ and defines an abstract neighborhood system there. Of course, not all functions in $F(X,P)$ will be bounded below. So we have to restrict ourselves to the subcone $B_l(X,P)$ of elements with this property, and $(B_l(X,P),V)$ then is a full locally convex cone.

Again, if (Q,V) is a locally convex cone, every subcone of $B_l(X,Q)$ is seen to be a locally convex cone as well. If, in particular, X is a topological space, we may consider the following subcones of $F(X,Q)$:

The cone	$C_u(X,Q)$	of functions continuous with respect to the upper topology on Q,
the cone	$C_l(X,Q)$	of functions continuous with respect to the lower topology on Q,
the cone	$C_s(X,Q)$	of functions continuous with respect to the symmetric topology on Q.

Their respective subcones of elements bounded below then are locally convex cones. If X is compact, then obviously all functions in $C_l(X,Q)$ and in $C_s(X,Q)$ are bounded below.

3. Local and global preorder. Closure.

Throughout this section we assume that Q is a locally convex cone, i.e. a subcone of the full locally convex cone (P,V). By means of the abstract neighborhood system V we shall define a new preorder on P, hence on Q, which in general will not coincide with the original one. It will however turn out to be more appropriate to describe the topological properties of a locally convex cone. We shall proceed in two steps:

3.1 The local preorder $\preceq_v$. For a fixed element $v \in V$ we define a relation $\preceq_v$ on P by:

$$a \preceq_v b \quad \text{if and only if} \quad a \le b+\rho v \quad \text{for all } \rho > 0.$$

It is easily seen that $\preceq_v$ is a preorder on P called the *v-local preorder* and that P endowed with this preorder and the abstract *0*-neighborhood system V is again a full locally convex cone. And again, as a subcone of P, Q is a locally convex cone. Clearly $a \le b$ in the original preorder implies $a \preceq_v b$.

3.2 The global preorder $\preceq$. The *global preorder* is defined as the intersection of all local preorders, i.e.

$$\begin{aligned} a \preceq b \quad &\text{iff} \quad a \le b+v \quad \text{for all } v \in V, \\ &\text{iff} \quad a \preceq_v b \quad \text{for all } v \in V. \end{aligned}$$

Endowed with this preorder and the abstract neighborhood system V, P is again a full locally convex cone, Q a locally convex cone, and $a \le b$ in the original preorder implies $a \preceq b$.

3.3 Closure. In Example 1.7 we studied decreasing convex subsets of a preordered cone and in particular the sets $\downarrow\{a\}$ generated by a single element $a \in P$. We shall do the same now with respect to the global preorder:

For every $a \in P$, the *closure of* a is defined to be the set

$$\begin{aligned} a &= \{b \in P \mid b \preceq a\} \\ &= \{b \in P \mid b \le a+v \quad \text{for all } v \in V\}. \\ &= \textstyle\bigcap_{v \in V} v(a). \end{aligned}$$

By definition, $\bar{a}$ is the intersection of all the upper neighborhoods of a. Clearly, $\bar{a}$ is convex and decreasing with respect to both of the preorders $\le$ and $\preceq$. For a subcone Q of P we may restrict the closure to Q and consider $\bar{a} \cap Q$ instead of $\bar{a}$.

We shall see in the following proposition that our notation of closure indeed refers to closure in a topological sense:

3.4 Proposition. *Let (Q,V) be a locally convex cone. For a subset $A \subset Q$ its closure $\overline{A}$ with respect to the lower topology is given by*

$$\overline{A} = \{b \in Q \mid \text{for all } v \in V \text{ there is } a \in A \text{ such that } b \le a+v\}.$$

In particular, $\overline{A}$ is decreasing with respect to the global preorder of Q, and convex if A is. In an analogous way the closure $\overline{\overline{A}}$ of A with respect to the upper topology is given by

$$\overline{\overline{A}} = \{b \in Q \mid \text{for all } v \in V \text{ there is } a \in A \text{ such that } a \le b+v\}.$$

Proof. We give the proof for the lower topology: Clearly $b \in \overline{A}$ if and only if $(b)v \cap A \neq \emptyset$ for all $v \in V$; i.e. there is some $a \in A$ such that $b \le a+v$. Now let $b \in \overline{A}$ and $c \lesssim b$. Then for $v \in V$, we have $c \le b+v/2$ and $b \le a+v/2$ for some $a \in A$, whence $c \le a+v$, and $c \in \overline{A}$ as well.

This renders immediately:

3.5 Corollary. *The closure $\overline{a}$ is the closure of $\{a\}$ with respect to the lower topology.*

3.6 Remark. (a) In a similar way, the set $\{b \mid a \lesssim b\}$, which is the intersection of all lower neighborhoods of a, is nothing but the closure of $\{a\}$ with respect to the upper topology. Finally, the set $\{b \mid b \lesssim a \text{ and } a \lesssim b\}$, which is the intersection of all symmetric neighborhoods of a, is only the closure of $\{a\}$ with respect to the symmetric topology.

(b) The above implies that the original preorder $\le$ and the global preorder $\lesssim$ on a locally convex cone coincide if and only if the sets $\downarrow\{a\} = \{b \mid b \le a\}$ are all closed in the lower topology, i.e. $\overline{a} = \downarrow\{a\}$.

The following observation turns out to be helpful in studying the locally convex cone $\overline{DConv(E)}$ for locally convex ordered vector spaces E.

3.7 Corollary. *If $Q = E$ is a locally convex ordered topological vector space (Example 2.7), $A \subset E$ a decreasing subset, then the closure of A in the lower and in the symmetric topologies coincide.*

Proof. Clearly, the symmetric closure of A is contained in the lower one. Now let b be in the closure with respect to the lower topology. Then for all $v \in V$ by Proposition 3.5 there is some $a \in A$ such that $b \le a+v$, i.e. $b \le a+w$ for an element w in the absolutely convex 0-neighborhood v of the vector space E. This shows that $a' = b-w \le a$ is an element of A as well, hence

$$b \le a'+v \quad \text{and} \quad a' < b+v,$$

and the symmetric neighborhood $v(b) \cap (b)v$ of b indeed meets A as well.

So $\overline{a}$ is in fact an element of $\overline{DConv(Q)}$, and a review of the definition of the addition $\overline{\oplus}$ there shows that $\overline{a} \,\overline{\oplus}\, \overline{b} = \overline{a+b}$. Thus, we have a canonical map

$$a \to \overline{a} : Q \to \overline{DConv(Q)}$$

which preserves the whole structure of Q. We denote by $\overline{Q}$ the image of Q under this map; i.e.

$$Q = \{\overline{a} \mid a \in Q\}.$$

As a subcone of $\overline{DConv(Q)}$ it is a locally convex cone by itself with the abstract 0-neighborhood system $\overline{V} = \{\overline{v} \mid v \in V\}$. The above mapping will however not be an embedding, i.e. one-to-one, in general.

3.8 Definition. A locally convex cone is called *separated* if $\overline{a} = \overline{b}$ implies $a = b$, i.e. if different elements have different closures.

This property corresponds to the Hausdorff axiom in locally convex vector spaces. The following is clear from the above:

3.9 Proposition. *For a locally convex cone (Q,V) the following properties are equivalent:*

- *(i)* *(Q,V) is separated.*
- *(ii)* *The upper topology on Q is T_0.*
- *(iii)* *The lower topology on Q is T_0.*
- *(iv)* *The symmetric topology on Q is T_0.*
- *(v)* *The symmetric topology on Q is Hausdorff.*

Moreover, on a separated cone the global preorder $\leq$ is in fact an order and the canonical map $a \to \overline{a}$ from Q with this order into $\overline{Q}$ is an isomorphism.

3.10 Corollary. *For every locally convex cone (Q,V), the cone $(\overline{Q},\overline{V})$ of one point closures is separated.*

We shall call $(\overline{Q},\overline{V})$ the *separated reflection* of (Q,V). This terminology is justified by the fact that continuous linear mappings from Q into separated locally convex cones all factor through $\overline{Q}$. This will be dealt with in detail in Chapter II. The following observation will be crucial for our further investigations in locally convex cones. It demonstrates the importance of the global preorder:

3.11 Proposition. *Let Q and Q' be locally convex cones. Then every mapping $f: Q \to Q'$, which is continuous with respect to the upper (or lower) topologies on both cones, is monotone with respect to their global preorders.*

Proof. Since a continuous mapping between topological spaces maps the closure of a subset into the closure of its image, this renders $f(\overline{a}) \subset \overline{f(a)}$ for all $a \in Q$, whenever f is continuous with respect to the lower topologies on Q and Q'. So clearly $b \leq a$, i.e. $b \in \overline{a}$ implies that $f(b) \in \overline{f(a)}$, hence $f(b) \leq f(a)$. If f is continuous for the upper topologies the same argument holds for the sets $\{b \mid a \leq b\}$.

3.12 Examples. Reviewing our standard Examples 2.6 through 2.9 for locally convex cones reveals the following:

(a) The cone $\overline{R}$ clearly is separated (Example 2.6).

(b) In Example 2.7 the locally convex ordered vector space $(E,\leq)$ is separated as a locally convex cone (E,V) as well, if only its positive cone E^+ is proper, i.e. $E^+\cap(-E^+) = \{0\}$: For elements $a,b \in E$, $a \preceq b$ translates into $b-a \in E^+ + V$, for all 0-neighborhoods $V \in \mathcal{V}$. Since E^+ is closed, this means $a \leq b$ in the given order of E, which therefore coincides with the global preorder. $\overline{a} = \overline{b}$ then implies $a = b$ because of the condition on E^+. Note that in the light of Proposition 3.11 this implies that mappings between locally convex ordered vector spaces, considered as locally convex cones, which are continuous with respect to their upper (or lower) topologies, need to be monotone with respect to the original orderings.

(c) For any locally convex cone $(Q,\mathcal{V})$ and $A,B \in Conv(Q)$, one has $A \preceq B$ if A is contained in the closure $\overline{A}$ of A with respect to the lower topology (c.f. Proposition 3.4). So in fact $(\overline{DConv(Q)}, \overline{\mathcal{V}})$ is the separated reflection of $(Conv(Q),\overline{\mathcal{V}})$ and, likewise, of $(DConv(Q),\overline{\mathcal{V}})$ (Example 2.8).

(d) Finally, in $(B_l(X,Q),\mathcal{V})$ (Example 2.9) for two Q-valued functions $f,g \in B_l(X,Q)$ we have $f \preceq g$ if $f(x) \preceq g(x)$ for all $x \in X$. So $B_l(X,Q)$ is a separated locally convex cone, whenever Q is.

4. Cancellation.

Let a,b,c be elements of an arbitrary preordered cone Q. In Section 1 we insisted on the fact that we do not require the cancellation law

(C) $\qquad a+c = b+c \quad$ implies $\quad a = b,$

as we wanted to include examples like the cone $Conv(E)$ of all non-empty convex subsets of a real vector space E. In this section we want to show that certain restricted cancellation laws hold in a locally convex cone $(Q,\mathcal{V})$. For the proofs, it is no restriction to assume this cone to be full. The first property holds in any preordered cone:

4.1 Lemma. $a+c \leq b+c$ *implies* $a+\rho c \leq b+\rho c$ *for all* $\rho \geq 0$.

Proof. Suppose $a+c \leq b+c$. Adding a and b, respectively, we obtain $2a+c \leq a+b+c$ and $a+b+c \leq 2b+c$, whence $2a+c \leq 2b+c$ or $a+c/2 \leq b+c/2$. Repeating the same argument, we obtain

$$a+\frac{c}{2^n} \leq b+\frac{c}{2^n}$$

for all natural numbers . If ρ is an arbitrary positive real number, choose an n such that $1/2^n \leq \rho$; adding $(\rho - 1/2^n)c$ to the last inequality, we obtain $a+\rho c \leq b+\rho c$ as desired.

If $c \geq 0$, then $a \leq a+\rho c$, and we may conclude:

4.2 Lemma. *If $c \geq 0$, then $a+c \leq b+c$ implies $a \leq b+\rho c$ for all $\rho > 0$.*

From now on we place ourselves in a locally convex preordered cone (Q,V):

4.3 Lemma. *Let c be v-bounded for some $v \in V$. Then*

$$a+c \leq b+c \text{ implies } a \leq b+\rho v \text{ for all } \rho > 0;$$

$$\text{i.e. } \quad a+c \leq b+c \text{ implies } a \lesssim_v b.$$

Proof. As c is v-bounded, there is a $\lambda > 0$ such that $c+\lambda v \geq 0$ and $c \leq \lambda v$. Suppose $a+c \leq b+c$. Then $a+(c+\lambda v) \leq b+(c+\lambda v)$. By Lemma 4.2 we conclude that $a \leq b+\rho(c+\lambda v) \leq b+2\rho\lambda v$ for all $\rho > 0$.

4.4 Proposition. *For every bounded element c one has :*

$$a+c \leq b+c \text{ implies } a+v \leq b+v \text{ for all } v \in V,$$

$$\text{i.e. } a+c \leq b+c \text{ implies } a \lesssim b, \text{ where } \lesssim \text{ is the global preorder.}$$

This is an immediate consequence of 4.3. We may apply this to the global preorder on Q, and we obtain the

4.5 Order cancellation for bounded elements:

$$a+c \lesssim b+c \text{ implies } a \lesssim b, \text{ whenever } c \text{ is bounded.}$$

For the closure of the elements we obtain (cf. Section 3):

4.6 Corollary. *Whenever c is bounded, one has:*

$$\bar{a}+\bar{c} \subset \bar{b}+\bar{c} \text{ implies } \bar{a} \subset \bar{b},$$

$$\bar{a}+\bar{c} = \bar{b}+\bar{c} \text{ implies } \bar{a} = \bar{b}.$$

As in separated cones the global preorder is an order, i.e. antisymmetric, we conclude:

4.7 Corollary. *For bounded elements c in a separated locally convex cone one has:*

$$a+c = b+c \text{ implies } a = b. \quad \text{(cancellation for bounded elements } c\text{).}$$

Now we are in a position to embed a separated locally convex cone (Q,V) with its global preorder into an extended locally convex preordered cone, in which the bounded elements form a vector space which is locally convex with respect to the symmetric topology:

4.8 Embedding. Let us consider first a full locally convex cone (P,V). By $\lesssim$ we denote the global preorder on P and by B the subcone of all bounded elements. In order to embed

B into a cone, in which the bounded elements become invertible, we perform the usual construction:

On the cone $P\times B$ of all pairs (a,b) with $a\in P$, $b\in B$, we define

$$(a,b)\leq(a',b') \text{ iff } a+b'\leq a'+b.$$

This relation on $P\times B$ clearly is reflexive. Let us verify that it is transitive as well: Let $(a,b)\leq(a',b')$ and $(a',b')\leq(a'',b'')$. Then $a+b'\leq a'+b$ and $a'+b''\leq a''+b'$. Adding b'' and b to these inequalities, respectively, we obtain

$$a+b'+b''\leq a'+b+b'' \text{ and } a'+b''+b\leq a''+b'+b,$$

whence $a+b'+b''\leq a''+b'+b$ by transitivity. As b' is bounded, we may cancel b' by 4.5, and we obtain $a+b''\leq a''+b$, i.e. $(a,b)\leq(a'',b'')$.

Thus, $\leq$ is a preorder on $P\mathrm{x}B$ which clearly is compatible with the addition and scalar multiplication. We have an obvious embedding

$$a\to(a,0)\ :\ P\to P\times B$$

which allows to consider P as a subcone of $P\times B$. Moreover, the set $V\times\{0\}$ of all pairs $(v,0)$, $v\in V$, is an abstract neighborhood system on the preordered cone $P\times B$. Using the cancellation property for bounded elements again, one proves that $(P\times B,V\times\{0\})$ is indeed a locally convex cone. We now pass to its separated reflection $\overline{P\times B}$ of one point closures $\overline{(a,b)}$ as in Section 3. For every bounded element b, the relations $(b,b)\leq(0,0)\leq(b,b)$ imply $\overline{(b,b)}=\overline{(0,0)}$. Thus $\overline{(b,0)}+\overline{(0,b)}=\overline{(b,b)}=\overline{(0,0)}$. We see that $\overline{(b,0)}$ has an additive inverse, namely $\overline{(0,b)}$. We may write shortly $\bar{a}$ for $\overline{(a,0)}$ and $-\bar{b}$ for $\overline{(0,b)}$, and $\bar{a}-\bar{b}$ for $\overline{(a,b)}$. Finally, we write $\overline{P\text{-}B}$ for $\overline{P\times B}$. If P is separated, the canonical map $a\to\bar{a}:P\to\bar{P}$ is an isomorphism with respect to the global preorder on P, and considering this, we may omit the bars: We have embedded (P,V), preserving its global preorder, into the locally convex cone $(P\text{-}B,V)$ in which the bounded elements form a vector space $E=B\text{-}B$. As we mentioned before (Section 2.5), the topology on E induced from the symmetric topology on the locally convex cone $(P\text{-}B,V)$ is given by neighborhoods of 0 in E:

$$v_0=(0)v\cap v(0)=\{b\in E\mid b\leq v \text{ and } 0\leq b+v\}$$
$$=\{b\in E\mid b\leq v \text{ and } -b\leq v\}.$$

For $v\in V$, these sets are convex and order convex and form the 0-neighborhood basis for a Hausdorff locally convex vector space topology on E. For an arbitrary element $c\in E$, the respective neighborhoods obviously are

$$v_c=(c)v\cap v(c)=c+v_0$$

Thus, the symmetric topology on $E=B\text{-}B$ is a Hausdorff locally convex vector space topology with the v_0, $v\in V$, as 0-neighborhood base. Note that the negative (hence also the positive) cone is closed in E by Lemma 3.5.

Until now, we have considered a full cone (P,V). For a subcone Q of P we may consider its set B_Q of bounded elements. Inside $P\text{-}B$ we form $Q\text{-}B_Q$ and we have embedded

Q in a cone where the bounded elements form a vector space $E_Q = B_Q - B_Q$. Thus we have proved:

4.9 Embedding Theorem. *Every separated locally convex cone* (Q,V) *can be embedded in a separated locally convex cone in which the bounded elements form a vector subspace* E_Q. *With respect to the symmetric topology* E_Q *is a locally convex ordered topological vector space. The embedding is faithfull for the abstract neighborhoods, whence for the global preorders.*

4.10 Remark. If we start with a Hausdorff locally convex topological vector space E and if we consider the cone $P = Conv(E)$ of all non-empty convex subsets of E, the cancellation laws 4.3 through 4.7 are well-known (c.f. Hörmander [26], Pinsker [42], Rådström [48], Rabinovich [46], [47], K. D. Schmidt [55]. Also the embedding of the cone $\overline{B}$ of non-empty closed bounded convex subsets into a locally convex topological vector space has been investigated by the same authors. It has been done in particular detail in [55].

5. Locally convex cones via convex quasiuniform structures.

We want to show that the notion of a locally convex cone can be defined in a natural way via quasiuniform structures in the sense of Nachbin [34]. For this, we use the notation $R \circ S = \{(a,c) \mid \text{there is } b \text{ such that } (a,b) \in R \text{ and } (b,c) \in S\}$ for the relational product for two binary relations R and S on a set Q, and R^{-1} for the converse relation $\{(b,a) \mid (a,b) \in R\}$; the diagonal is denoted by $\triangle$. We recall the following:

5.1 Quasiuniform structures. A collection $\mathcal{U}$ of subsets of $Q \times Q$ is called a *quasiuniform structure on* Q, if the following hold:

(U1) $\triangle \subset U$ for every $U \in \mathcal{U}$;

(U2) for all $U,V \in \mathcal{U}$ there is a $W \in \mathcal{U}$ such that $W \subset U \cap V$;

(U3) for all $U \in \mathcal{U}$ there is a $V \in \mathcal{U}$ such that $V \circ V \subset U$.

Nachbin has called this a basis of a quasiuniform structure, as it requires $\mathcal{U}$ to be a filter basis only and not a filter.

To every quasiuniform structure $\mathcal{U}$ on Q we associate

(a) a preorder defined by $a \leq b$ iff $(a,b) \in U$ for all $U \in \mathcal{U}$; the graph of this preorder is the set $\gamma = \bigcap_{U \in \mathcal{U}} U$.

(b) two topologies: The neighborhood bases for an element a for the ***upper*** and *lower* topology are given by the sets

$$U(a) = \{b \in Q \mid (b,a) \in U\},\ U \in \mathcal{U},$$
$$(a)U = \{b \in Q \mid (a,b) \in U\},\ U \in \mathcal{U},$$

respectively.

(c) a *uniform structure* $\mathcal{U}_s = \{U \cap U^{-1} \mid U \in \mathcal{U}\}$; the topology associated with this uniform structure is the common refinement of the lower and upper topology; it is Hausdorff iff the preorder $\preceq$ is in fact an order.

5.2 The quasiuniform structure of a locally convex cone. Let Q be a preordered cone and $\mathcal{V}$ an abstract neighborhood system (contained in some preordered cone $P \supset Q$). For every abstract neighborhood $v \in \mathcal{V}$, we put

$$\tilde{v} = \{(a,b) \in Q \times Q \mid a \leq b+v\}$$

The collection $\tilde{\mathcal{V}}$ of all the $\tilde{v}$, $v \in \mathcal{V}$, is a quasiuniform structure:
As $0 \leq v$, we have $a \leq a+v$ for all a, whence (U1). As $\mathcal{V}$ is directed downward, we also have (U2). Furthermore, we have the property

(U3') $\quad (\lambda v)^\sim \circ (\mu v)^\sim \subset ((\lambda+\mu)v)^\sim$ for all $\lambda,\mu > 0$;

indeed, $a \leq b+\lambda v$ and $b \leq c+\mu v$ imply $a \leq c+(\lambda+\mu)v$. Choosing $w = v/2$ in (U'3) we obtain $\tilde{w} \circ \tilde{w} \subset \tilde{v}$, whence (U3).
In addition, it is easy to see that all the $\tilde{v}$ are convex subsets of $Q \times Q$, and

(U4) $\quad$ for $\tilde{v} \in \tilde{\mathcal{V}}$, $\lambda > 0$, we have $\lambda\tilde{v} \in \tilde{\mathcal{V}}$ as well.

Finally, if $(Q,\mathcal{V})$ is a locally convex cone, the condition that every element has to be bounded below, translates into:

(U5) $\quad$ For all $a \in Q$ and $\tilde{v} \in \tilde{\mathcal{V}}$ there is some $\rho > 0$ such that $(0,a) \in \rho\tilde{v}$.

So we have a convex quasiuniform structure in the following sense:

5.3 Convex quasiuniform structures. Let Q be a cone. A collection $\mathcal{U}$ of convex subsets $U \subset Q \times Q$ is called a *convex quasiuniform structure*, if $\mathcal{U}$ satisfies the properties (U1), (U2),

(U3') $\quad (\lambda U) \circ (\mu U) \subset (\lambda+\mu)U$ for all $U \in \mathcal{U}$ and $\lambda,\mu > 0$,

and (U4) $\quad \lambda U \in \mathcal{U}$ for all $U \in \mathcal{U}$ and $\lambda > 0$.

If we start with an abstract neighborhood system $\mathcal{V}$ on a preordered cone as above, the lower and upper topologies on Q are precisely the lower and upper topologies associated with the quasiuniform structure $\tilde{\mathcal{V}}$. The preorder associated with this quasiuniform structure coincides with the global preorder on Q; indeed, $(a,b) \in \tilde{v}$ for every $v \in \mathcal{V}$ means $a \leq b+v$ for every $v \in \mathcal{V}$. Thus, the global preorder and the various topologies on a locally convex preordered cone are all described by the quasiuniform structure $\tilde{\mathcal{V}}$.

5.4 The abstract neighborhood system for a convex quasiuniform structure. Let $\mathcal{U}$ be a convex quasiuniform structure on a cone Q. We shall embed Q in a preordered cone

$\boldsymbol{P}$ containig an abstract neighborhood system $\boldsymbol{V}$ in such a way that the canonical quasiuniform structure $\tilde{\boldsymbol{V}}$ associated with $\boldsymbol{V}$ is equivalent to the original quasiuniform structure $\boldsymbol{U}$ on $\boldsymbol{Q}$.

For this, let $\boldsymbol{B}$ be any 'subbasis' of $\boldsymbol{U}$, which means that for every $U \in \boldsymbol{U}$ one can find $U_1,...,U_n \in \boldsymbol{B}$ and $\lambda_1,...,\lambda_n > 0$ such that $\lambda_1 U_1 \cap ... \cap \lambda_n U_n \subset U$. Let $\boldsymbol{V}$ be the set of all families $r = (r_U)_{U \in \boldsymbol{B}}$, where r_U is a strictly positive real number for finitely many $U \in \boldsymbol{B}$ and $r_U = +\infty$ else.

Adjoining a zero (0) to $\boldsymbol{V}$, we obtain an ordered cone $\boldsymbol{V_0}$ with componentwise defined operations and order.

Let $\boldsymbol{P}$ be the direct sum $\boldsymbol{P} = \boldsymbol{Q} \oplus \boldsymbol{V_0}$ with the usual addition and scalar multiplication. Define a preorder on $\boldsymbol{P}$ in the following way:

$$x \oplus r \le y \oplus s \text{ if } r \le s \text{ and } (x,y) \in \lambda U \text{ for all } \lambda > s_U - r_U \text{ whenever } s_U < +\infty.$$

This relation clearly is reflexive. Let us show transitivity: Let $x \oplus r \le y \oplus s \le z \oplus t$. Then firstly $r \le s \le t$; secondly, consider any U with $t_U < +\infty$, and let $\lambda > t_U - r_U$. Then there are λ_1, λ_2 such that $\lambda = \lambda_1 + \lambda_2$, $\lambda_1 > s_U - r_U$, $\lambda_2 > t_U - s_U$. We conclude that $(x,y) \in \lambda_1 U$ and $(y,z) \in \lambda_2 U$, whence $(x,z) \in \lambda_1 U \circ \lambda_2 U \subset (\lambda_1 + \lambda_2) U = \lambda U$ by (U3'). Thus $x \oplus r \le z \oplus t$.

Let us show the compatibility of this preorder with the algebraic operations: Let $x \oplus r \le y \oplus s$. Clearly, $\lambda(x \oplus r) \le \lambda(y \oplus s)$. Let us show that $(x \oplus r) + (z \oplus t) \le (y \oplus s) + (z \oplus t)$: Firstly, $r \le s$ implies $r+t \le s+t$. Secondly, take any U such that $s_U + t_U < +\infty$. Then r_U, s_U, t_U are all finite and for every $\lambda > (s_U + t_U) - (r_U + t_U) = s_U - r_U$ we have $(x,y) \in \lambda U$. As $(z,z) \in \varepsilon U$ for all $\varepsilon > 0$ we conclude that $(x+z, y+z) \in (\lambda + \varepsilon) U$, whence the assertion.

Now we have proved that $\boldsymbol{P} = \boldsymbol{Q} \oplus \boldsymbol{V_0}$ is a preordered cone. It is easy to see that $\boldsymbol{V}$ (identified with $\{0\} \oplus \boldsymbol{V}$) is an abstract neighborhood system on $\boldsymbol{P}$.

When is $x \le y \oplus r$ for $x,y \in \boldsymbol{Q}$ and $r \in \boldsymbol{V}$? Let $U_1,...,U_n$ be the members of $\boldsymbol{B}$ such that r_U is finite. Then

$$x \le y \oplus r \text{ iff } (x,y) \le \lambda_i U_i \text{ for all } \lambda_i > r_{U_i} \text{ and all } i,$$

i.e. $$x \le y \oplus r \text{ iff } (x,y) \in \lambda (r_{U_1} U_1 \cap ... \cap r_{U_n} U_n) \text{ for all } \lambda > 1.$$

As $\boldsymbol{B}$ was chosen to be a 'subbasis' of the quasiuniform structure $\boldsymbol{U}$, we see that $\boldsymbol{U}$ is equivalent to the quasiuniform structure $\tilde{\boldsymbol{V}}$ consisting of all $\tilde{r} = \{(x,y) \mid x \le y \oplus r\}$.

We summarize:

5.5 Proposition. *The notions of an abstract neighborhood system* $\boldsymbol{V}$ *and a convex quasiuniform structure* $\boldsymbol{U}$ *for a cone* $\boldsymbol{Q}$ *are equivalent in the following sense:*

For every abstract neighborhood system $\boldsymbol{V}$ *for a preordered cone* $\boldsymbol{Q}$ *there is a convex quasiuniform structure* $\tilde{\boldsymbol{V}}$ *on* $\boldsymbol{Q}$ *which induces the global preorder on* $\boldsymbol{Q}$ *and the same upper, lower and symmetric topologies.*

If $\boldsymbol{Q}$ *is a cone with a convex quasiuniform structure* $\boldsymbol{U}$, *then one can find a preorder and an abstract neighborhood system* $\boldsymbol{V}$ *for* $\boldsymbol{Q}$ *such that the quasiuniform structure* $\tilde{\boldsymbol{V}}$ *is equivalent to* $\boldsymbol{U}$.

As we mentioned above, the condition of lower boundedness for elements of a locally convex cone translates into condition (U5) in terms of the convex quasiuniform structure. Thus, the two approaches to locally convex cones firstly via a preorder and an abstract neighborhood system and secondly via convex quasiuniform structures turned out to be equivalent. In fact, in applications the second approach will often arise more naturally, as it avoids the explicit construction of a full cone containing the abstract neighborhood system.

5.6 Defining locally convex structures through quasimetrics. In the same sense that a locally convex vector space topology can be defined through a family of seminorms, a locally convex structure on a cone can also be defined through a family of sublinear quasimetrics. Here, we just want to give a short hint to this fact without going into details:

Let C be a cone. A function $d : C\times C \to \overline{\mathbb{R}}_+$ is called a *quasimetric*, if

(M1) $d(a,a) = 0$ for all $a \in C$

and (M2) $d(a,c) \leq d(a,b) + d(b,c)$ for all $a,b,c \in C$.

Note that the symmetry $d(a,b) = d(b,a)$ is not required. A quasimetric is called *sublinear*, if it satisfies (M3) $d(a+a', b+b') \leq d(a,b)+d(a',b')$ for all $a,a',b,b' \in C$

and (M4) $d(\alpha a, \alpha b) = \alpha\, d(a,b)$ for all $a,b \in C$ and $\alpha > 0$.

Although we allow infinite quasidistances, we shall restrict our attention to quasimetrics satisfying (M5) $d(0,a) \leq +\infty$ for all $a \in C$.

Note that in the presence of (M1) and (M2) the subadditivity (M3) can be replaced by the axiom of subinvariance

(M3') $d(a+c, b+c) \leq d(a,b)$ for all $a,b,c \in C$.

We now consider a family $(d_i)_{i\in I}$ of sublinear quasimetrics. Such a family is called *directed*, if for any $i,j \in I$, there is a $k \in I$ and a $\lambda > 0$ such that

$$\max\{d_i(a,b), d_j(a,b)\} \leq \lambda d_k(a,b) \quad \text{for all } a,b \in C.$$

5.7 Proposition. *Giving a directed family* $(d_i)_{i\in I}$ *of sublinear quasimetrics on a cone* C *is equivalent to giving a convex quasiuniform structure and, hence, by Proposition 5.5, to a locally convex structure.*

Indeed, a convex quasiuniform structure $\mathcal{U}$ gives rise to a directed family $(d_U)_{U\in\mathcal{U}}$ of sublinear quasimetrics by defining

$$d_U(a,b) = \inf\{\rho > 0 \mid (a,b) \in \rho U\} \quad \text{for all } U \in \mathcal{U} \text{ and } a,b \in C.$$

Conversely, if $(d_i)_{i\in I}$ is a directed family of sublinear quasimetrics, we define

$$U_{i,\rho} = \{(a,b) \in C\times C \mid d_i(a,b) < \rho\} \quad \text{for all } i \in I \text{ and all } \rho > 0.$$

Thus we obtain a convex quasiuniform structure. The axiom (M5) for quasimetrics corresponds to the axiom (U5) for quasiuniform structures. The details are left to the reader.

Chapter II: Uniformly Continuous Operators and the Dual Cone

In this chapter we develop the basic functional analysis for locally convex cones as far as we shall need it for our applications. After the basic definition of uniform continuity for linear operators on locally convex cones in Section 1 we turn to linear functionals in Section 2, and we prove our basic Hahn-Banach type theorems, e.g. the Sandwich Theorem 2.8, the Extension Theorem 2.9, the Separation Theorems 2.10 and 2.14. In Section 3 we present some rudiments for a duality theory for locally convex cones culminating in a Mackey-Arens type result (3.8). In Section 4 we turn to compact convex subsets C of the dual cone Q^* and prove some results on extreme points and faces thereof. All these results are analogues of standard theorems for locally convex vector spaces. But the proofs cannot be transferred directly to our more general situation in cones.

The last two sections of this chapter have a more special flavor. In Section 5 we introduce a class of cones which correspond to locally convex vector lattices (Riesz spaces) with an M-structure. The linear operators that correspond to lattice homomorphisms in the vector lattice case are called directional operators in Section 6. These notions allow to identify the extreme points of the polars v_Q° of 0-neighborhoods v in Q (see Theorem 6.7).

At a first reading one can skip Sections 3 to 6.

1. Uniformly continuous operators.

1.1 Linear operators. For cones Q and P, a map $T:Q\to P$ is called a *linear operator*, if

$$T(a+b) = T(a)+T(b) \quad \text{for all } a,b\in Q \text{ and}$$
$$T(\alpha a) = \alpha T(a) \quad \text{for all } a\in Q \text{ and } \alpha\geq 0.$$

Note that this implies $T(0_Q)=0_P$.

1.2 Uniformly continuous mappings. In the following let (Q,V) and (P,W) be two preordered locally convex cones. A mapping $T:Q\to P$ is called *uniformly continuous* or *u-continuous* for short, if for every $w\in W$ one can find a $v\in V$ such that

$$a\leq b+v \quad \text{implies} \quad T(a)\leq T(b)+w.$$

Uniform continuity is not just continuity. It is immediate from the definition that it implies and combines continuity with respect to the upper, lower and symmetric topologies on Q and P.

1.3 Remarks. (a) Let us consider on Q and P the quasiuniform structures $\tilde{V}$ and $\tilde{W}$ as in Ch. I, section 5, i.e. $\tilde{V}$ is the collection of all

$$\tilde{v} = \{(a,b) \in Q\times Q \mid a \le b+v\}, \quad v \in V$$

and likewise for $\tilde{W}$. Then a mapping $T: Q \to P$ is u-continuous if and only if it is uniformly continuous with respect to these quasiuniform structures in the sense that for every $\tilde{w} \in \tilde{W}$ there is a $\tilde{v} \in \tilde{V}$ such that $(a,b) \in \tilde{v}$ implies $(T(a),T(b)) \in \tilde{w}$.

(b) If Q is a full cone, then a monotone linear operator $T: Q \to P$ is u-continuous, if for every $w \in W$ there is a $v \in V$ such that $T(v) \le w$. Indeed, if $T(v) \le w$, then $a \le b+v$ implies $T(a) \le T(b+v) = T(b)+T(v) \le T(b)+w$.

As an immediate consequence from Proposition 3.11 in Ch.I we have:

1.4 Monotonicity Lemma. *Let $T: Q \to P$ be a u-continuous mapping. Then $a \precsim b$ implies $T(a) \precsim T(b)$, where $\precsim$ denotes the global preorder on Q and P, respectively.*

Since for the original preorder $\le$ on Q, $a \le b$ implies $a \precsim b$ we conclude that $a \le b$ implies $T(a) \precsim T(b)$ as well.
Using the notations of Ch. I, sec. 3 and 4, we obtain:

1.5 Proposition. *Let $T: Q \to P$ be u-continuous and linear. There is a unique corresponding u-continuous linear operator*

$$\overline{T}: \overline{Q}\text{-}\overline{B}_Q \to \overline{P}\text{-}\overline{B}_P$$

such that $\overline{T}(\overline{a}) = \overline{T(a)}$ for all $a \in Q$, where B_Q and B_P denote the subcones of bounded elements in Q and P, respectively, and $\overline{a}$ and $\overline{T(a)}$ the closures of a and $T(a)$. Moreover, if P is separated, this correspondence is onto and one-to-one, and u-continuous linear operators from Q into P and from $\overline{Q}\text{-}\overline{B}_Q$ into $\overline{P}\text{-}\overline{B}_P$ may be identified.

Proof. Firstly, we observe that under T the image of each bounded element of Q is bounded in P: Indeed, let $a \in B_Q$ and $w \in W$. Then there is $v \in V$ such that $a \le b+v$ implies $T(a) \le T(b)+w$. As $a \le \rho v$ for some $\rho > 0$, this shows $T(a) \le \rho w$. Secondly, if $\overline{a} = \overline{b}$ for $a,b \in Q$, then $T(a) \precsim T(b)$ as well as $T(b) \precsim T(a)$. Thus, $\overline{T(a)} = \overline{T(b)}$, and $\overline{T}$ is well defined; u-continuity and linearity on $\overline{Q}\text{-}\overline{B}_Q$ are easily checked. Thirdly, any linear operator $\overline{T}: \overline{Q} \to \overline{P}$ has a unique extension from $\overline{Q}\text{-}\overline{B}_Q$ into $\overline{P}\text{-}\overline{B}_P$. Finally, if P is separated, then for each operator $\overline{T}$ from $\overline{Q}\text{-}\overline{B}_Q$ into $\overline{P}\text{-}\overline{B}_P$ the formula $T(a) = b$, whenever $\overline{T}(\overline{a}) = \overline{b}$, defines an operator from Q into P ($\overline{c} = \overline{b}$ implies $c = b$ then). So, indeed, both sets of operators may be identified.

1.6 Examples. (a) Addition $(a,b) \to a+b$ is a u-continuous operator from $Q\times Q$ into Q. Indeed, given $w \in V$, we choose $v = w/2$, and we obtain that $a \le c+v$ and $b \le d+v$ imply $a+b \le c+d+w$.

(b) For fixed $\lambda \geq 0$, the operator $a \to \lambda a : Q \to Q$ is u-continuous.

(c) For every bounded element $b \geq 0$, the map $\lambda \to \lambda b : R_+ \to Q$ is u-continuous. Indeed, given $w \in V$, choose an $\varepsilon > 0$ such that $b \leq \varepsilon w$. Then $\mu \leq \lambda + \varepsilon$ and $b \geq 0$ imply $\mu(b) \leq (\lambda+\varepsilon)b = \lambda b + \varepsilon b \leq \lambda b + w$.

(d) As an immediate consequence of (a) and (b) we obtain: For finitely many bounded positive elements $b_1,\dots,b_n$ in Q, the map

$$(\lambda_1,\dots,\lambda_n) \to \sum_i \lambda_i b_i \ : \ R_+^n \to Q$$

is u-continuous.

(e) Let E and F be locally convex topological vector spaces and $T : E \to F$ a continuous linear operator. We extend T to non-empty convex subsets by defining $\overline{T}(A) = \{T(a) \mid a \in A\}$. In this way we obtain a linear operator

$$\overline{T} : Conv(E) \to Conv(F)$$

which is readily verified to be u-continuous.

The same procedure can be applied more generally to extend any u-continuous operator $T : Q \to P$ between arbitrary locally convex cones to a u-continuous linear operator $\overline{T} : Conv(Q) \to Conv(P)$. Extension to the cones $DConv(Q)$ and $\overline{DConv(Q)}$ (for their algebraic and locally convex structures, see Ch. I, Example 2.8), however, requires a slight modification in the definition of $\overline{T}$:
For $A \in DConv(Q)$, respectively $A \in \overline{DConv(Q)}$, we set

$$\overline{T}(A) = \downarrow\{T(A)\} \in DConv(P), \quad \text{respectively} \quad \overline{T}(A) = \overline{T(A)} \in \overline{DConv(P)},$$

where in the latter case closure is meant with respect to the lower topology on P. In order to check the linearity of $\overline{T}$, we have to recall the addition as defined in those cones. Let us do this in the latter case: For $A, B \in \overline{DConv(Q)}$ we have $A \overline{\oplus} B = \overline{A+B}$, whence

$$\overline{T}(A \overline{\oplus} B) = \overline{T(\overline{A+B})} = \overline{T(A+B)} = \overline{T(A)+T(B)} = \overline{\overline{T}(A)+\overline{T}(B)} = \overline{T}(A) \overline{\oplus} \overline{T}(B).$$

(The second equality comes from the fact that $T(\overline{A+B}) \subset \overline{T(A+B)}$, as T is continuous with respect to the lower topologies on Q and P.) Finally, since $\overline{\alpha A} = \alpha \overline{A}$ (this follows easily looking at Proposition 3.4 in Ch. I), we have $\overline{T}(\alpha A) = \alpha \overline{T}(A)$ as well; u-continuity is immediately checked.

Reviewing Proposition 1.5 we now see that the operator $T : Q \to P$ even may be extended to an operator

$$\overline{T} : \overline{DConv(Q)}\text{-}\overline{B}_Q \to \overline{DConv(P)}\text{-}\overline{B}_P.$$

($\overline{Q}$ and $\overline{P}$ are subcones of $\overline{DConv(Q)}$ and $\overline{DConv(P)}$, respectively.) This extension of T, however, will not be unique.

1.7 The cone of u-continuous linear operators. For two linear operators S and T from Q into P and for $\lambda \geq 0$, the sum $S+T$ and λT are also linear. If both S and T are u-continuous, so are $S+T$ and λT by 1.6(a,b). Thus, the u-continuous linear operators from Q into P form a cone $L(Q,P)$. We do not intend to discuss locally convex structures for this cone in this paper.

2. Linear functionals and the separation theorems.

Throughout this section, let (Q,V) be a locally convex cone with its preorder $\leq$ and the global preorder $\preceq$. A *linear functional* on Q is a linear operator $\mu : Q \to \overline{R}$.

2.1 The dual cone. According to 1.1, a linear functional μ on Q is called *uniformly continuous* or simply *u-continuous*, if there is a $v \in V$ such that

(O) $\qquad a \leq b+v$ implies $\mu(a) \leq \mu(b)+1$.

Every u-continuous linear functional is monotone by 1.4, even with respect to the global preorder on Q. The u-continuous linear functionals on Q form again a cone (c.f. 1.7), denoted by Q^* and called the *dual cone* of Q. We shall endow Q^* with the topology $w(Q^*,Q)$ of pointwise convergence of the elements of Q, considered as functions on Q with values in $\overline{R}$ with its usual topology. (Studying duality theory, later on we shall consider $\overline{R}$ with its symmetric topology, which isolates $+\infty$, and the resulting finer topology $s(Q^*,Q)$ on Q^* as well.)

If Q is a full cone and μ a monotone linear functional on Q, then the definition of u-continuity becomes particularly simple: μ will be u-continuous if and only if there is a $v \in V$ such that $\mu(v) \leq 1$. For linear functionals Proposition 1.5 yields:

2.2 Proposition. *Q^* and $(\overline{Q\text{-}B_Q})^*$ may be identified.*

2.3 Polars. For every $v \in V$, the *polar* of v is defined to be the set v_Q° of all linear functionals μ on Q satisfying (O), i.e.

$$v_Q^\circ = \{\mu \in Q^* \mid a \leq b+v \text{ implies } \mu(a) \leq \mu(b)+1\}$$

We simply write v° instead of v_Q°, if no confusion is possible. If Q is a full cone, the definition of the polar becomes particularly simple: v_Q° is the set of all monotone linear functionals on Q such that $\mu(v) \leq 1$.

2.4 Proposition. *The polar v° of any $v \in V$ is a compact convex subset of Q^* in the topology $w(Q^*,Q)$.*

Proof. Clearly, v° is convex and closed for pointwise convergence in the set of all functions $f : Q \to \overline{R}$. Remember that every $a \in Q$ is bounded below, i.e. there is a real number $\lambda_a > 0$ such that $0 \leq a+\lambda_a v$. For every μ in v° we then have $0 = \mu(0) \leq \mu(a)+\lambda_a$, whence $\mu(a) \geq -\lambda_a$. Thus, v° is in fact a closed subset of the compact space $\prod_{a\in Q} [-\lambda_a,+\infty]$ and hence compact.

It is an important feature of our notion of locally convex cones that with only slight additional requirements we are able to prove all of the desired Hahn-Banach type theorems on the existence of sufficiently many u-continuous linear functionals. Our prime model *Conv(E)* for locally convex cones has plenty of them, indeed: Every continuous linear functional μ on a locally convex topological vector space E may be extended by defining

$$\overline{\mu}(A) = \sup\{ \mu(a) \mid a \in A \} \quad \text{for every non-empty convex subset } A \text{ of } E.$$

Thus, we obtain a functional $\overline{\mu} : Conv(E) \to \overline{\mathbf{R}}$ which is easily verified to be linear and u-continuous. For deriving our Hahn-Banach type theorems in general we cannot apply directly the corresponding results that one finds e.g. in the book of Fuchssteiner and Lusky [22]; the reason is that in the literature mostly linear and sublinear functionals with values in $\mathbf{R} \cup \{-\infty\}$ are considered, whilst we have to deal with functionals which have values in $\overline{\mathbf{R}} = \mathbf{R} \cup \{+\infty\}$. The difference between those two points of view is essential and not just a question of reversing the order on $\mathbf{R}$. We have to begin with a few remarks on sublinear functionals:

2.5 Sublinear functionals. A map $p : \mathbf{Q} \to \overline{\mathbf{R}}$ is *sublinear* if

$$p(a+b) \le p(a)+p(b) \quad \text{and} \quad p(\lambda a) = \lambda p(a) \quad \text{for all } a,b \in \mathbf{Q} \text{ and } \lambda \ge 0.$$

According to 1.2 a sublinear functional p is called *u-continuous*, if there is a neighborhood $v \in V$ such that

(O) $\quad a \le b+v$ implies $p(a) \le p(b)+1$ whenever $a,b \in \mathbf{Q}$.

Every u-continuous sublinear functional is monotone by Proposition 1.4, even monotone with respect to the global preorder on $\mathbf{Q}$.

If $\mathbf{Q}$ is a full cone and p a monotone sublinear functional on $\mathbf{Q}$, then the definition of u-continuity becomes particularly simple: p will be u-continuous if and only if there is a $v \in V$ such that $p(v) \le 1$.

2.6 Lemma. *Let* $\mathbf{Q}$ *be a subcone of the preordered locally convex cone* $(\mathbf{P},\mathbf{V})$. *Every u-continuous sublinear functional on* $\mathbf{Q}$ *can be extended to a u-continuous sublinear functional on* $\mathbf{P}$*; more precisely: Let* p *be a sublinear functional on* $\mathbf{Q}$ *and* $v \in V$ *such that*

(O) $\quad a \le b+v$ *implies* $p(a) \le p(b)+1$, *whenever* $a,b \in \mathbf{Q}$,

then there is a sublinear functional $\overline{p}$ *on* $\mathbf{P}$ *extending* p *such that*

(O') $\quad x \le y+v$ *implies* $\overline{p}(x) \le \overline{p}(y)+1$, *whenever* $x,y \in \mathbf{P}$.

Proof. For every $x \in \mathbf{P}$ we define

$$\overline{p}(x) = \inf\{ p(a)+\lambda \mid x \le a+\lambda v \text{ for some } a \in \mathbf{Q} \text{ and } \lambda > 0 \}.$$

We verify:

(*i*) $\overline{p}$ is an extension of p: Let $x \in \mathbf{Q}$. Clearly, $\overline{p}(x) \le p(x)$. For the converse inequality consider any α such that $\overline{p}(x) < \alpha$. Then there is an $a \in \mathbf{Q}$ and a $\lambda > 0$ such that $x \le a+\lambda v$ and $p(a)+\lambda < \alpha$. Condition (O) implies $p(x) \le p(a)+\lambda < \alpha$. As this holds for every $\alpha > \overline{p}(x)$, we infer $p(x) \le \overline{p}(x)$.

(*ii*) $\overline{p}$ satisfies (O'): Let $x,y \in \mathbf{P}$ with $x \le y+v$. We want to show that $\overline{p}(x) \le \overline{p}(y)+1$. This is clear if $\overline{p}(y) = +\infty$. So suppose that $\overline{p}(y) < +\infty$. For every

$\alpha > \tilde{p}(y)$ we may find $a \in Q$ and $\lambda>0$ such that $y \le a+\lambda v$ and $p(a)+\lambda \le \alpha$. We conclude that $x \le y+v \le a+(\lambda+1)v$, whence $\tilde{p}(x) \le p(a)+\lambda+1 < \alpha+1$. As this holds for every $\alpha > \tilde{p}(y)$, we obtain $\tilde{p}(x) \le \tilde{p}(y)+1$.

(*iii*) $\tilde{p}$ is sublinear: The proofs are similar to the above, and we leave them to the reader.

2.7 Superlinear functionals. A map $q : Q \to \mathbf{R} \cup \{+\infty,-\infty\}$ is *superlinear* if

$$q(\lambda a) = \lambda q(a) \text{ for all } \lambda \ge 0,\ a \in Q \text{ and}$$

$$q(a+b) \ge q(a)+q(b) \text{ for all } a,b \in Q \text{ such that both } q(a) \text{ and } q(b) \text{ are finite.}$$

2.8 Sandwich Theorem. *Let (Q,V) be a preordered locally convex cone. Let $p : Q \to \overline{\mathbf{R}}$ be sublinear and $q : Q \to \mathbf{R} \cup \{+\infty,-\infty\}$ superlinear with $q(a) \le p(a)$ for all $a \in Q$. If p is u-continuous, then there is a u-continuous linear functional $\mu : Q \to \overline{\mathbf{R}}$ such that*

$$q(a) \le \mu(a) \le p(a) \quad \text{for all } a \in Q;$$

more precisely: If, for some $v \in V$, the functional p satisfies

(O) $\quad a \le b+v$ *implies* $p(a) \le p(b)+1$ *whenever* $a,b \in Q$,

then there is a linear functional μ in the polar v_Q° of v such that

$$q(a) \le \mu(a) \le p(a) \quad \text{for all } a \in Q.$$

Proof. We may suppose that Q is a full cone. In fact, by Lemma 2.6 we can extend p to a full cone containig Q without disturbing hypothesis (O), and q may be extended by defining its value to be $-\infty$ on the new elements. Now we consider the subset of Q

$$Q_f = \{ a \in Q \mid p(a) < +\infty \}.$$

As p is sublinear and monotone, Q_f is a decreasing subcone of Q. Let us show that Q_f is a face of Q: Let a,b be elements of Q such that $a+b \in Q_f$, i.e. $p(a+b) < +\infty$. Recall that in a locally convex cone every element is bounded below. Hence there is a $\lambda > 0$ such that $0 \le a+\lambda v$, whence $b \le a+b+\lambda v$. As p satisfies (O), we infer that $p(b) \le p(a+b)+\lambda$, whence $p(b) < +\infty$, i.e. $b \in Q_f$.

On Q_f the sublinear functional p does not take the value $+\infty$, and we may apply the Theorem 1.2.5 in [22] which assures the existence of a monotone linear functional μ on Q_f with values in $\mathbf{R} \cup \{-\infty\}$ such that $q(a) \le \mu(a) \le p(a)$ for all $a \in Q_f$. By (O) we have $p(v) \le 1$, whence $\mu(v) \le 1$. This shows that in fact μ does not attain the value $-\infty$: As every $a \in Q$ is bounded below, there is $\lambda_a > 0$ such that $0 \le a+\lambda_a v$; i.e. $0 = \mu(0) \le \mu(a)+\lambda_a$ and $\mu(a) \ge -\lambda_a$. We extend μ to Q by $\mu(a) = +\infty$ for all $a \notin Q_f$. As Q_f is a decreasing face, μ is still monotone and linear, and obviously $q(a) \le \mu(a) \le p(a)$ holds for all $a \in Q$. Finally, $\mu(v) \le 1$ implies $\mu \in v_Q^\circ$.

2.9 Extension Theorem. *Let Q be a subcone of the locally convex cone (P,V). Then every u-continuous linear functional on Q can be extended to a u-continuous linear functional on P; more precisely: For every $\mu \in v_Q^\circ$ there is a $\tilde{\mu} \in v_P^\circ$ such that $\mu = \tilde{\mu}|_Q$.*

Proof. By Lemma 2.6 we may extend μ to a sublinear functional p on P which satisfies (O) for all $a,b \in P$. We define a superlinear functional q on P by

$$q(a) = \begin{cases} \mu(a), & \text{if } a \in Q, \\ -\infty, & \text{else.} \end{cases}$$

By the Sandwich Theorem 2.8 we find a linear functional $\tilde{\mu} \in v_Q^\circ$ such that $q(a) \le \tilde{\mu}(a) \le p(a)$ for all $a \in P$. As $q(a) = \mu(a) = p(a)$ for all $a \in Q$, the functional $\tilde{\mu}$ extends μ.

2.10 Weak separation. *Let $a,b \in Q$ with $b \ge 0$ and let $v \in V$ such that $a \not\le b+v$. Then there is a linear functional μ on Q such that*

$$\mu(a) \ge 1 \quad \text{and} \quad \mu(x) \le 1 \ \text{whenever } x \le b+v;$$

more generally:
$$\mu(x) \le \mu(y)+1 \quad \text{whenever } x \le y+b+v.$$

Proof. Let $w = b+v$. We may suppose that $w \in V$, as $b \ge 0$. Let $Q_o = \{\alpha a \mid \alpha \ge 0\}$ and define $\mu_0 : Q_o \to \overline{R}$ by

$$\mu_0(\alpha a) = \alpha \inf\{\lambda > 0 \mid a \le \lambda w\}.$$

Then μ_0 is a linear functional on Q_o contained in the polar of w in Q_o^*. By the Extension Theorem 2.9, μ_0 has a linear extension μ to Q which is contained in the polar of w in Q^*. The extension μ has the desired properties.

In order to obtain strict separation, we have to reinforce the hypothesis:

2.11 Lemma. *Let $v \in V$, and let a be a v-bounded and b an arbitrary element of Q such that $a \not\le b+\rho v$ for some $\rho > 1$. Then there is a linear functional $\mu \in v_Q^\circ$ such that $\mu(a) > \mu(b)+1$.*

Proof. We may suppose that Q is a full cone. First we choose an α such that $0 \le b+\alpha v$. We then have

$$a+\alpha v \not\le b+\alpha v+\sigma v \quad \text{whenever } 1<\sigma<\rho.$$

Indeed, the inequality $a+\alpha v \not\le b+\alpha v+\sigma v$ would imply $a \le b+\rho v$ for all $\rho > \sigma$ by the Cancellation Lemma 4.2, Ch. I. Replacing a by $a+\alpha v$ and b by $b+\alpha v$ in 2.10, we may find a linear functional $\nu \in Q^*$ such that

(i) $\quad \nu(a+\alpha v) \ge 1$ $\quad$ and $\quad$ (ii) $\quad \nu(b+\alpha v+\sigma v) \le 1$.

From these inequalities we conclude that

$$\nu(a)+\alpha\nu(v) \ge \nu(b)+\alpha\nu(v)+\sigma\nu(v), \quad \text{whence}$$

(iii) $\quad \nu(a) \ge \nu(b)+\sigma\nu(v)$.

From (ii) we also get that $\nu(v) \ne +\infty$. Indeed, if $\nu(v) = 0$, then we choose $\lambda > 0$ such that $a \le \lambda v$ (which is possible, as a is supposed to be v-bounded) and we get $\nu(a+\alpha v) \le \nu(\lambda v+\alpha v) = 0$ which contradicts (i). So (iii) implies

(iv) $\quad \nu(a) > \nu(b)+\nu(v)$.

Now we put $\mu = \nu/\nu(v)$. Then (iv) becomes $\mu(a) > \mu(b)+1$, and μ is contained in the polar of v as $\mu(v) = 1$.

2.12 Strict separation. The separation property, as formulated in the preceding lemma, will in general not hold for unbounded elements a in an arbitrary locally convex cone (Q,V). This property, however, turns out to be crucial in the investigation of Korovkin type approximation.

We shall say that (Q,V) has the *strict separation property*, if the following holds:

(SP) For all $a,b \in Q$ and $v \in V$ such that $a \not\leq_v b+v$, i.e $a \not\leq b+\rho v$ for some $\rho > 1$, there is a linear functional $\mu \in v_Q^\circ$ such that $\mu(a) > \mu(b)+1$.

(Note that $\mu(x) \leq \mu(y)+1$, whenever $x \leq y+v$, as $\mu \in v_Q^\circ$.) In view of the preceding lemma, we will obtain strict separation if we have sufficiently many bounded elements in Q. This will be guaranteed by a property which we shall introduce now and which may be easily verified for our prime examples.

2.13 Tight coverage by bounded elements. A locally convex cone (Q,V) is said to be *tightly covered by its bounded elements* if for all $a,b \in Q$ and $v \in V$ such that $a \notin v(b)$ there is some bounded element $a' \in Q$ such that $a' \leq a$ and $a' \notin v(b)$.

This renders the

2.14 Separation Theorem. *Every locally convex cone* (Q,V) *which is tightly covered by its bounded elements has the strict separation property* (SP).

Proof. By (2.13) choose a bounded element $a' \leq a$ such that $a' \not\leq b+\rho' v$ for some $\rho' > 1$. Lemma 2.11 yields a linear functional $\mu \in v_Q^\circ$ such that $\mu(a') \geq \mu(b)+\rho'$. As $\mu(a) \geq \mu(a')$, we have the desired result.

2.15 Adjoint operators. For locally convex cones (Q,V) and (P,W) and a u-continuous linear operator $T: Q \to P$ we define *the adjoint operator of* T by

$$(T^*(\nu))(a) = \nu(T(a)) \quad \text{for all } \nu \in P^* \text{ and } a \in Q.$$

Clearly, $T^*(\nu) \in Q^*$, and T^* is a linear operator from P^* into Q^*; more precisely: If for $v \in V$ and $w \in W$ we have

$$a \leq b+v \quad \text{implies} \quad T(a) \leq T(b)+w$$

then T^* maps w° into v°: Indeed, if $\nu \in w^\circ$ and $a \leq b+v$, then

$$(T^*(\nu))(a) = \nu(T(a)) \leq \nu(T(b))+1 = (T^*(\nu))(b)+1.$$

2.16 Examples. Firstly, we note that $\overline{Q}\text{-}\overline{B}_Q$ has the separation property whenever Q has: $(\bar{a}\text{-}\bar{b}) \not\leq (\bar{c}\text{-}\bar{d})+v$ means $(\overline{a+d}) \not\leq (\overline{c+b})+v$. So there is $\mu \in Q^* = (\overline{Q}\text{-}\overline{B}_Q)^*$ such that $\mu \in v_Q^\circ$ and $\mu(a+d) > \mu(c+d)$, i.e. $\mu(\bar{a}\text{-}\bar{b}) > \mu(\bar{c}\text{-}\bar{d})$.

Now let us investigate the dual cones of our standard examples from the previous chapter:

For all $\rho \geq 0$, clearly $\overline{R}^*$ contains the mappings $a \rightarrow \rho a$, $a \in \overline{R}$. (Recall that $0 \cdot \infty = 0$.) If we add the mapping 0^* which maps all $a \in R$ into 0 and $+\infty$ into $+\infty$, then we have all of $\overline{R}^*$. Clearly $\overline{R}$ has property (SP).

The dual cone of the locally convex ordered vector space $(E,\leq)$ (Ch. I, Example 2.7) consists exactly of the positive functionals in the dual space of E. Again, (SP) is obvious, as $(E,\leq)$ satisfies 2.13.

If (Q,V) is a locally convex cone, then the dual cones of $Conv(Q)$, $DConv(Q)$ and $\overline{DConv(Q)}$ all coincide, since $\overline{DConv(Q)}$ is the separate reflection of $Conv(Q)$. With $\mu \in Q^*$ we associate an element $\overline{\mu} \in Conv(Q)^*$ defined by

$$\overline{\mu}(A) = \sup\{\mu(a) \mid a \in A\} \quad \text{for every } A \in Conv(Q).$$

(Clearly $\overline{\mu}$ belongs to the polar of the same neighborhood as μ.) But of course, not all elements of $Conv(Q)^*$ are of this type.

If B denotes the subcone of the bounded elements of Q, and D is any a subcone of $Conv(B)$, which contains all singleton sets $\{b\}, b \in B$, then property 2.13, whence the separation property clearly holds for D:

Suppose $A \not\leq A' + \overline{v}$ for sets $A, A' \in D$, and $v \in V$, i.e. there is some $a \in A$, such that $a \not\leq a' + v$ for all $a' \in A'$. Since a is bounded in Q, so is $\{a\}$ in D, and we have $\{a\} \leq A$, but $\{a\} \not\leq A' + \overline{v}$.

If X is a set and (Q,V) is a locally convex cone, then for each $x \in X$ and $\mu \in Q^*$ we obtain a linear functional μ_x in the dual cone of $(B_l(X,Q),V)$ (Ch. I, Example 2.9) by $\mu_x(f) = \mu(f(x))$ for all $f \in B_l(X,Q)$. Of course, not all elements of $B_l(X,Q)^*$ will be of this type, but it is already clear from the above that the separation property holds for $B_l(X,Q)$ if it holds for Q.

2.17 Example: Normed spaces. In the context of studying Korovkin type approximation for linear contractions on normed spaces the following locally convex cone and its dual will be of interest: Let $(E,\|\ \|)$ be a normed vector space with unit ball B. Let Q be the subcone of $Conv(E)$ consisting of all sets $a+\rho B$, $a \in E$, $\rho \geq 0$. Then the dual cone of Q may be easily described: It is the set of all $\mu \oplus r$, where $r \geq 0$ and μ is a linear functional on E such that $\|\mu\| \leq r$, if we define $(\mu \oplus r)(a+\rho B) = \mu(a) + r\rho$. The polar of B (as an abstract neighborhood for Q) consists of those $\mu \oplus r$ with $r \leq 1$.

2.18 Example: Compact convex subsets of R^n. No straightforward description seems to be available for the dual cone of $Conv(R^n)$. However, using a representation technique for compact convex subsets of R^n as developed in [28] we obtain a handy representation for the dual cone of $CConv(R^n)$, the cone of all non-empty compact convex subsets of R^n: By B we denote the unit ball of R^n, by Y the dual unit sphere of B which is compact and topologically isomorphic to the Euclidean unit sphere S^{n-1}. With each $A \in CConv(R^n)$ we associate the classical support functional p_A on Y, which is defined by

$$p_A(y) = \sup \{y(a) \mid a \in A\}.$$

p_A is easily seen to be continuous on Y (for details c.f. [28]). The mapping

$$A \to p_A \; : \; CConv(R^n) \to C(Y)$$

is an embedding, as it is easy to check that for sets $A,C \in CConv(R^n)$ and $\varepsilon > 0$ we have

$$A \subset C+\varepsilon B \text{ if and only if } p_A \leq p_C+\varepsilon.$$

By the Stone-Weierstraß Theorem (the supremum of p_A and p_C is just the support function for the convex hull of the sets A and C) the linear subspace generated by the functions p_A is dense in $C(Y)$. So the dual cone of $CConv(R^n)$ may be identified with the dual cone of $C(Y)$, i.e. the cone of all positive Radon measures on Y.

2.19 Example: $\overline{R}$-valued functions. Let X be any a set and Y a covering of X, i.e. a collection of subsets whose union is all of X. Consider the cone P of $\overline{R}$-valued functions which are bounded below on every $Y \in Y$, endowed with its usual (pointwise) order. For each $Y \in Y$ and $\varepsilon > 0$ we define the function $\psi_{Y,\varepsilon}$ on X by:

$$\psi_{Y,\varepsilon}(x) = \begin{cases} \varepsilon, & \text{if } x \in Y, \\ +\infty, & \text{if } x \notin Y. \end{cases}$$

Then the cone V generated by those functions defines an abstract neighborhood system on P which thus is seen to become a full locally convex cone. Its bounded elements are nothing but the functions on X which are bounded on every $Y \in Y$. Property 2.13 is easily checked for P. So the strict separation property holds for all subcones of P.

Now, let (Q,V) be any separated locally convex cone with the strict separation property. If we set $X = Q^*$ and $Y = \{v^\circ \mid v \in V\}$, then the elements of Q may be considered as $\overline{R}$-valued functions on X. For $a,b \in Q$ and $v \in V$ we have by (SP)

$$a \leq b+v \quad \text{if and only if as functions on } X \quad a \leq b+\psi_{v^\circ,1}.$$

So Q has a representation as a locally convex cone of $\overline{R}$-valued functions.

This gives rise to a complete characterization of separated locally convex cones with separation property:

2.20 Representation Theorem. *For a separated locally convex cone* (Q,V) *the following properties are equivalent:*

(i) (Q,V) *has the strict separation property.*

(ii) (Q,V) *may be represented as a cone of* $\overline{R}$*-valued functions on some set* X *in the sense of Example 2.19.*

(iii) (Q,V) *may be represented as a subcone of* $\overline{DConv(E)}$ *for some locally convex ordered vector space* E.

Representation is meant to be a one-to-one embedding which preserves the algebraic structure and the abstract neighborhoods of (Q,V).

Proof. The implication (i)$\Rightarrow$(ii) was shown above. (iii)$\Rightarrow$(i) is an immediate consequence of the Separation Theorem 2.14, as $\overline{DConv(E)}$ carries property 2.13 by Example 2.16, whence any subcone has the strict separation property as well. It remains to show that (ii) implies (iii):

Let Q be a cone of $\overline{R}$-valued functions on the set X in the sense of Example 2.19, Y the respective collection of subsets of X, and V the abstract neighborhood system induced by Y. Now let E be the vector space of all R-valued functions on X which are bounded on every $Y \in Y$, endowed with its canonical order and the symmetric topology induced by V. This is the locally convex space which we will use.

For each $a \in Q$ now consider the subset of E:

$$C(a) = \{ b \in E \mid b \leq a \}.$$

The set $C(a)$ clearly is convex and decreasing. Recall that for decreasing subsets of E the closure with respect to the lower and the symmetric topologies coincides (Ch I, Proposition 3.7). $C(a)$ is closed with respect to the lower as well as the symmetric topology. Thus, we have a mapping $a \rightarrow C(a)$ of Q into the cone $\overline{DConv(E)}$ of closed convex and decreasing subsets of E. Recall that $\overline{DConv(E)}$ is a locally convex cone with its addition $\overline{\oplus}$ and the abstract neighborhoods via those in E, as described in Ch I, Example 2.8. The above mapping clearly is one-to-one and faithful with respect to the algebraic operations:

Recall that $C(a)\overline{\oplus}C(a') = \overline{C(a)+C(a')}$. As $b \leq a$ and $b' \leq a'$ implies $b+b' \leq a+a'$, this renders immediately $C(a)\overline{\oplus}C(a') \subset C(a+a')$ for all $a,a' \in Q$. If on the other hand $b \leq a+a'$ for some $b \in E$ then we choose any function $c \in E$, such that $c \leq a'$ (for example, $c = a' \wedge 0$, the pointwise infimum of a' and 0). Now we set $d = a \wedge (b-c)$ and $d' = b-d$. We easily check that $d,d' \in E$, $d \leq a$ and $d' \leq a'$. So indeed, $b \in C(a)+C(a') \subset C(a)\overline{\oplus}C(a')$.

Finally, let $Y \in Y$, $\varepsilon > 0$, and $\psi_{Y,\varepsilon}(x)$ as in Example 2.19. By $v \in V$ we denote the corresponding neighborhood in Q which via E induces a neighborhood $\overline{v}$ on $\overline{DConv(E)}$ as well. For $a,a' \in Q$ we have $a \in v(a')$ iff $a \leq a'+\psi_{Y,\varepsilon}$, and $C(a) \in \overline{v}(C(a')$ iff for every function $b \in E$ such that $b \leq a$ there is some $b' \in E$, such that $b' \leq a'$ and $b \leq b'+\psi_{Y,\varepsilon}$. Both conditions are immediately seen to be equivalent. So our embedding indeed preserves the abstract neighborhoods as well.

If all elements of the locally convex cone (Q,V) are bounded then they are bounded below with respect to the symmetric topology. Thus, the symmetric quasiuniform structure in the sense of Ch. I.5 defines a locally convex cone topology on Q as well. Let us denote this by $(Q,\overline{V})$; i.e. for $a,b \in Q$, $v \in V$ we have

$$a \leq b+\overline{v} \text{ if and only if } a \leq b+v \text{ and } b \leq a+v.$$

(Recall form Ch. I, 4.8 that Q may be embedded into a vector space which is locally convex in this topology.) We shall conclude this section with the description of the dual cone of $(Q,\overline{V})$ by means of the dual cone of (Q,V) :

2.21 Proposition. *Suppose that all elements of the locally convex cone* $(\mathcal{Q},\mathcal{V})$ *are bounded. A linear mapping*

$$\mu : \mathcal{Q} \to \mathbf{R}$$

is uniformly continuous with respect to the symmetric topology on $\mathcal{Q}$ *if and only if it has a representation as the difference of two elements of the dual cone* $\mathcal{Q}^*$ *of* $\mathcal{Q}$*; i.e. the dual of* $\mathcal{Q}$ *for the symmetric topology is given by* $\mathcal{Q}^*\text{-}\mathcal{Q}^*$.

Proof. It is obvious that for $\mu \in \mathcal{Q}^*$, since its values on $\mathcal{Q}$ are all finite, both μ and $-\mu$ are continuous with respect to the symmetric topology. So differences of the elements of $\mathcal{Q}^*$ indeed render such linear mappings.

If on the other hand $\mu : \mathcal{Q} \to \mathbf{R}$ is linear and u-continuous with respect to the symmetric topology, then there is some $v \in \mathcal{V}$ such that

$$a \le b+v \text{ and } b \le a+v \text{ imply } |\mu(a)-\mu(b)| \le 1.$$

On $\mathcal{Q}\times\mathcal{Q}$ we introduce the abstract neighborhoods $\bar{v}$ $(v \in \mathcal{V})$ by

$$(a,b) \le (a',b')+\bar{v} \quad \text{if and only if} \quad a \le a'+v \text{ and } b' \le b+v.$$

As the elements of $\mathcal{Q}\times\mathcal{Q}$ are all bounded, $(\mathcal{Q}\times\mathcal{Q}, \bar{\mathcal{V}})$ thus becomes a locally convex cone. On the subcone $\Lambda = \{(a,a) \mid a \in \mathcal{Q}\}$ of $\mathcal{Q}\times\mathcal{Q}$ the linear functional

$$\tilde{\mu} : \Lambda \to \mathbf{R}, \quad \text{defined by} \quad \tilde{\mu}(a,a) = \mu(a),$$

then clearly is u-continuous ($(a,a) \le (b,b)+\bar{v}$ means $a \le b+v$ and $b \le a+v$). By the Extension Theorem 2.9 it has a u-continuous extension $\tilde{\nu}$ to all of $\mathcal{Q}\times\mathcal{Q}$.
Now for $a \in \mathcal{Q}$ we set $\mu_1(a) = \tilde{\nu}(a,0)$ and $\mu_2(a) = -\tilde{\nu}(0,a)$. Because $a \le b+v$ translates into $(a,0) \le (b,0)+\bar{v}$ and $(0,b) \le (0,a)+\bar{v}$, this implies

$$\mu_1(a) = \tilde{\nu}(a,0) \le \tilde{\nu}(b,0)+1 = \mu_1(b)+1 \quad \text{and} \quad \mu_2(a) = -\tilde{\nu}(0,a) \le -\tilde{\nu}(0,b)+1 = \mu_2(b)+1.$$

So indeed, both μ_1 and μ_2 are elements of $\mathcal{Q}^*$, and $\mu = \mu_1\text{-}\mu_2$.

3. Duality theory for locally convex cones.

As in duality theory for locally convex topological vector spaces we shall study dual pairs of cones and locally convex structures induced by polar topologies. This will render another approach to locally convex cones with separation property (SP) and their duals, including a Mackey-Arens type characterization of topologies compatible with this duality.

3.1 Dual pairs. A *dual pair* $(\mathcal{Q},\mathcal{P})$ consists of two cones $\mathcal{Q}$ and $\mathcal{P}$ together with a bilinear map (i.e. a map which is linear in both variables)

$$(a,x) \to \langle a,x\rangle : \mathcal{Q}\times\mathcal{P} \to \overline{\mathbf{R}}.$$

Obvious examples are: A cone $\mathcal{Q}$ and the cone $\mathcal{P}$ of all linear mappings on $\mathcal{Q}$ into $\overline{\mathbf{R}}$ with the evaluation as the bilinear form on $\mathcal{Q}\times\mathcal{P}$, and for a locally convex cone $(\mathcal{Q},\mathcal{V})$ the pair $(\mathcal{Q},\mathcal{Q}^*)$, where $\mathcal{Q}^*$ denotes the dual cone of $\mathcal{Q}$.

3.2 Polar topologies. Let (Q,P) be a dual pair. A subset X of P is said to be *σ-bounded below*, if for all $a \in Q$ we have

$$\inf \{\langle a,x\rangle \mid x \in X\} > -\infty.$$

Now let $\mathcal{X}$ be a collection of subsets of P which are σ-bounded below and fulfill the properties

(P1) $\lambda X \in \mathcal{X}$ for all $X \in \mathcal{X}$ and $\lambda > 0$.

(P2) For all $X,Y \in \mathcal{X}$ there is some $Z \in \mathcal{X}$ such that $X \cup Y \subset Z$.

Using this we define a convex quasiuniform structure on Q in the sense of Ch. I, 5: For $X \in \mathcal{X}$ let U_X be the convex subset of $Q \times Q$

$$U_X = \{(a,b) \in Q \times Q \mid \langle a,x\rangle \leq \langle b,x\rangle + 1 \text{ for all } x \in X\}.$$

It is evident that conditions (U1), (U2), (U3') and (U4) from Ch. I, 5 hold. Moreover, since the sets $X \in \mathcal{X}$ all are σ-bounded below, this shows (U5), and the collection $\mathcal{U}_{\mathcal{X}}$ of all such sets U_X thus defines a locally convex structure on Q. We shall refer to it as the *$\mathcal{X}$-topology* on Q.

For a set $X \in \mathcal{X}$ we shall denote by v_X the abstract neighborhood induced on Q by U_X. As usual, v_X° is the polar of v_X, i.e. the set of all linear functionals $\mu : Q \to \overline{R}$ such that for $a,b \in Q$

$$\langle a,x\rangle \leq \langle b,x\rangle + 1 \text{ for all } x \in X \quad \text{implies} \quad \mu(a) \leq \mu(b) + 1.$$

Clearly, if we consider the elements of P as linear functionals on Q, then $x \in v_X^\circ$ for all $x \in X$ and the subcone generated by $\bigcup_{X \in \mathcal{X}} X$ is contained in Q^*. This shows in particular that the strict separation property (SP) holds for any $\mathcal{X}$-topology on Q: If $(a,b) \notin U_X$, then there is some $x \in X \subset v_X^\circ$ such that $x(a) > x(b) + 1$.

3.3 Examples. (a) Let $\mathcal{X}$ be the set of all finite subsets of P. The resulting $\mathcal{X}$-topology on Q will be called the *weak*-topology* $\sigma(Q,P)$. One obtains $\sigma(P,Q)$ in an analogous way.

(b) Let $(Q,\mathcal{V})$ be a locally convex cone with separation property (SP). Consider the dual pair (Q,Q^*) and the collection $\mathcal{X}$ of subsets of Q^* which consists of the polars of the abstract neighborhoods $v \in \mathcal{V}$. Then clearly $\mathcal{X}$ satisfies (P1) and (P2), and the resulting $\mathcal{X}$-topology on Q is equivalent to the original one: If for $a,b \in Q$ and $v \in \mathcal{V}$ we have $a \leq b+v$, this implies $\mu(a) \leq \mu(b)+1$ for all $\mu \in v^\circ$, whence $(a,b) \in U_{v^\circ}$. If on the other hand $a \not\leq b+2v$, then by (SP) there is a linear functional $\mu \in v^\circ$ such that $\mu(a) > \mu(b)+1$, whence $(a,b) \notin U_{v^\circ}$. This shows in particular that every locally convex topology on a cone with (SP) may be considered as a polar topology. The subsets of Q^* defining it were all seen to be $w(Q^*,Q)$-compact (Proposition 2.4).

3.4 The symmetric topologies $s(P,Q)$ and $w(P,Q)$. We are going to study now polar topologies on Q which are generated by subsets of P with certain topological properties. To do so, we firstly investigate two special topologies on P: By $s(P,Q)$ we denote the symmetric topology induced on P by $\sigma(P,Q)$. A typical neighborhood for an element $x \in P$ in this topology is given via a finite subset $A = \{a_1,...,a_n\}$ of Q by

$$S_A(x) = \left\{ y \in P \;\middle|\; \begin{array}{ll} |\langle a_i,y\rangle - \langle a_i,x\rangle| \leq 1, & \text{if } \langle a_i,x\rangle < +\infty \\ \langle a_i,y\rangle = +\infty, & \text{if } \langle a_i,x\rangle = +\infty \end{array} \right\}.$$

In general, this topology is finer than $w(P,Q)$, the topology of pointwise convergence of the elements of Q considered as functions on P with values in $\overline{R}$. For $P = Q^*$ this topology was introduced in 2.1. A typical neighborhood for $x \in P$ in $w(P,Q)$ is given by (again $A = \{a_1,...,a_n\}$ is a finite subset of Q):

$$W_A(x) = \left\{ y \in P \;\middle|\; \begin{array}{l} |\langle a_i,y\rangle - \langle a_i,x\rangle| \leq 1, \text{ if } \langle a_i,x\rangle < +\infty \\ \langle a_i,y\rangle \geq 1, \text{ if } \langle a_i,x\rangle = +\infty \end{array} \right\}.$$

For $a \in Q$ let 0_a be the linear functional on P defined by

$$0_a(x) = \begin{cases} 0, & \text{if } \langle a,x\rangle < +\infty, \\ +\infty, & \text{if } \langle a,x\rangle = +\infty. \end{cases}$$

Considering the elements of Q as linear functionals on P, we derive for the topologies $s(P,Q)$ and $w(P,Q)$:

3.5 Proposition. *If for every $a \in Q$ we have $0_a \in Q$ as well, then the topologies $w(P,Q)$ and $s(P,Q)$ on P coincide.*

Proof. For $x \in P$, every neighborhood $S_A(x)$ as defined in 3.5 coincides with some $W_B(x)$. This is obvious, if we let $B = A \cup \{0_a \mid a \in A\}$.

Note that this applies in particular to a locally convex cone P with its dual P^*: We have $0_\mu \in P^*$ for all $\mu \in P^*$, whence the topologies $w(P,P^*)$ and $s(P,P^*)$ coincide.

3.6 Lemma. *Let X be an $s(P,Q)$-compact subset of P, and suppose that $\mu \in v_X^\circ$ attains only finite values on Q. Then for $Y = \{x \in X \mid \langle a,x\rangle < +\infty$ for all $a \in Q\}$, we have $\mu \in v_Y^\circ$ as well.*

Proof. For a finite subset $A = \{a_1,...,a_n\}$ of Q let

$$X_A = \{ x \in X \mid \langle a_i,x\rangle < +\infty \text{ for all } i = 1,..,n \}.$$

X_A then is $s(P,Q)$-compact as well, since the subsets $\{ x \in P \mid \langle a_i,x\rangle = +\infty \}$ are all open in the topology $s(P,Q)$. We shall check first that we have $\mu \in v_{X_A}^\circ$ as well: Indeed, let $a,b \in Q$ such that $(a,b) \in U_{X_A}$, i.e. $\langle a,x\rangle \leq \langle b,x\rangle + 1$ for all $x \in X_A$. Then this implies for the elements

$$a_0 = a + \sum_{i=1}^{n} a_i \quad \text{and} \quad b_0 = b + \sum_{i=1}^{n} a_i$$

that $\langle a_0,x\rangle \leq \langle b_0,x\rangle + 1$ holds even for all $x \in X$. Using $\mu \in v_X^\circ$ we conclude $\mu(a_0) \leq \mu(b_0)+1$, and as the values of μ are all finite, $\mu(a) \leq \mu(b)+1$ as well.

Now let $\mathcal{A}$ be the set of all finite subsets of Q. Then clearly $Y = \bigcap_{A\in\mathcal{A}} X_A$, and Y is $s(P,Q)$-compact as well.
If, contradicting our claim, we had $\mu \notin v_Y^\circ$ then we could find elements $a,b \in Q$ such that $\langle a,x\rangle \le \langle b,x\rangle+1$ for all $x \in Y$, but $\mu(a) > \mu(b)+\rho$ for some $\rho > 1$. The element μ was seen to be in $v_{X_A}^\circ$ for all $A \in \mathcal{A}$, so in each of the sets X_A we find an element x_A such that $\langle a,x_A\rangle > \langle b,x_A\rangle+\rho$. The set $\mathcal{A}$ is ordered and directed upward by inclusion, so we may consider the net $(x_A)_{A\in\mathcal{A}}$ which by the compactness of X may in turn be supposed to be convergent with limit point $x \in X$. This shows $\langle a,x\rangle \ge \langle b,x\rangle+\rho$, and the special type of the neighborhoods in $s(P,Q)$ guarantees that the value $\langle b,x\rangle$ still remains finite, as all the values $\langle b,x_A\rangle$ are: Indeed, if we had $\langle b,x\rangle = +\infty$ then this would imply $\langle b,y\rangle = +\infty$ for all elements y in a suitable neighborhood $S_B(x)$ as defined in 3.5 (set $B = \{b\}$), whence $\langle b,x_A\rangle = +\infty$ for all $A \in \mathcal{A}$ larger than some A_0, contradicting the above construction of the net $(x_A)_{A\in\mathcal{A}}$. But this shows that x cannot be an element of Y (otherwise we had $\langle a,x\rangle \le \langle b,x\rangle+1$). So there is some set $A \in \mathcal{A}$ such that $x \notin X_A$.
Finally, $V = P\backslash X_A$ is a neighborhood of x, so there is some $A' \ge A$ such that for all $A'' \ge A'$ we have $x_{A''} \in V$. But this contradicts $x_{A''} \in X_{A''} \subset X_A$ and completes our proof.

3.7 Lemma. *Let (Q,P) be a dual pair such that its bilinear form attains only finite values. Let X be an $s(P,Q)$-compact convex subset of P containing 0_P, and suppose that $\mu \in v_X^\circ$ attains only finite values on Q. Then there is $x \in X$ such that*

$$\mu(a) = \langle a,x\rangle \quad \textit{for all } a \in Q.$$

Proof. Let E be the vector space of all linear mappings $f : Q \to \mathbb{R}$ endowed with the topology of pointwise convergence. Then E is a locally convex topological vector space, and X may be embedded (not necessarily one-to-one) into E as a compact convex subset. In this sense Q-Q may be considered as a subspace of the continuous affine functions on X, and since $\mu \in v_X^\circ$ we have:

$$\mu(h) \le 1 \text{ for all } h \in Q\text{-}Q \text{ such that } \sup\{h(x) \mid x \in X\} \le 1.$$

By a well-known result about compact convex sets (c.f. Alfsen's book [1]) then μ as a mapping on Q-Q coincides with a positive multiple of a point evaluation on X; more precisely: There are $y \in X$ and $\lambda \in [0,1]$ such that

$$\mu(a) = \lambda\langle a,y\rangle \quad \text{for all } a \in Q.$$

Because $0 \in X$ by assumption and X is convex, we have $x = \lambda y \in X$ as well; this completes our proof.

Now we are ready to formulate a Mackey-Arens type result for locally convex cones:

3.8 Theorem. *Let (Q,P) be a dual pair and $X \subset P$ the union of finitely many $s(P,Q)$-compact convex subsets of P. Then for every $\mu \in v_X^\circ$, i.e. every linear functional $\mu : Q \to \overline{R}$ such that for $a,b \in Q$*

$$\langle a,x\rangle \leq \langle b,x\rangle + 1 \quad \textit{for all } x \in X \quad \textit{implies} \quad \mu(a) \leq \mu(b)+1,$$

there is an element $x \in P$ such that

$$\mu(a) = \langle a,x\rangle \quad \textit{for all } a \in Q \textit{ with } \mu(a) < +\infty.$$

Proof. Let $Q' = \{a \in Q \mid \mu(a) < +\infty\}$ and $Y = \{x \in X \mid \langle a,x\rangle < +\infty \text{ for all } a \in Q'\}$. Note that Y is an $s(P,Q)$-closed subset of X, as for all $a \in Q$ the sets $\{x \in X \mid \langle a,x\rangle < +\infty\}$ are closed in this particular topology. So Y is the union of finitely many $s(P,Q)$-compact convex subsets of P as well. Following Lemma 3.6, then we conclude that $\mu \in v_Y^\circ$. Now we set $P' = \{\lambda y \mid y \in Y'\}$ and apply Lemma 3.7 to the dual pair (Q',P'): Clearly, our bilinear mapping is finite on $Q' \times P'$, and the subset Y of P' is $s(P',Q')$-compact, as this topology is coarser than $s(P,Q)$. The same holds for the convex hull Z of $Y \cup \{0\}$, and we may apply Lemma 3.7 with Z in place of X. The element $x \in Z$ from this lemma then has the desired property.

3.9 Remark. (a) Note that the above theorem applies to the weak*-topology $\sigma(Q,P)$, which is generated by the finite subsets X of P.

(b) If the bilinear mapping on (P,Q) attains the value $+\infty$, then the convex hull of finitely many $s(P,Q)$-compact convex subsets of P need not be $s(P,Q)$-compact, as we shall show in the following example: Let $P = \overline{Conv(R)}$, i.e. the cone of all closed intervals in R, $Q = R_+^2$, endowed with the bilinear mapping

$$\langle [a,b], (x,y)\rangle = by - ax \quad \text{and} \quad \langle [a,+\infty), (x,y)\rangle = \begin{cases} -ax, & \text{if } y = 0, \\ +\infty, & \text{if } y > 0. \end{cases}$$

For the elements $[1,+\infty)$ and $\{0\}$ in P, i.e. the singleton subsets containing them, their convex hull consists of all intervals $[\lambda,+\infty)$, $0 < \lambda \leq 1$, and of $\{0\}$. But this subset of P clearly fails to be $s(P,Q)$-compact, as it does not contain $[0,+\infty) \in P$ which is an element of its closure.

In locally convex vector spaces the closure of a convex subset may be described by means of continuous linear functionals, i.e. is an invariant for all topologies of the same duality. An similar result does not seem to be available for locally convex cones with unbounded elements. Another such invariant, however, as it is known in vector spaces, the notion of boundedness for subsets, is preserved in locally convex cones as well. This is an implication of the following:

3.10 Theorem. *Let (Q,V) be a locally convex cone and $p : Q \to \overline{R}$ a u-continuous sublinear functional. If p is unbounded (as a function) on the subset A of Q then there is a linear functional $\mu \in Q^*$ such that $\mu(a) \leq p(a)$ for all $a \in Q$, and μ is unbounded on A as well.*

Proof. By Lemma 2.6 we may assume that $\mathcal{Q}$ is a full cone. So any monotone linear functional on $\mathcal{Q}$ which is dominated by p is u-continuous.

If $\inf\{p(a) \mid a \in A\} = -\infty$ then the same holds for every linear functional dominated by p and our claim is obvious.

If there is an element $a \in A$ such that $p(a) = +\infty$ then it follows immediately from the Sandwich-Theorem that there is $\mu \in \mathcal{Q}^*$ such that $\mu \leq p$ and $\mu(a) = +\infty$ as well.

So we only have to consider the case that p is finite on A and that $\sup\{p(a) \mid a \in A\} = +\infty$. Moreover (eventually after restricting to the respective subcone), we may assume that p is finite on all of $\mathcal{Q}$. Now let S be the subset of all $\mu \in \mathcal{Q}^*$ which are dominated by p. S is non-empty by the Sandwich-Theorem, $w(\mathcal{Q}^*,\mathcal{Q})$-compact as a closed subset of some polar $v^\circ_{\mathcal{Q}}$ and convex. For each $n \in N$ now set

$$S_n = \{\mu \in S \mid \sup\{|\mu(a)| \mid a \in A\} \leq n\}.$$

If $S \neq \bigcup_{n \in N} S_n$, this proves our claim. Otherwise, as S is a Baire space, there is some S_n with non-empty topological interior; i.e. we have some $\mu_0 \in S_n$ and elements $a_1, a_2, \dots a_m \in \mathcal{Q}$ such that the set

$$U = \{\mu \in S \mid |\mu(a_i) - \mu_0(a_i)| \leq 1, \ i = 1,2,\dots m\}$$

is contained in S_n. (As p is finite on $\mathcal{Q}$, this holds for all $\mu \in S$ as well, so U is indeed a general neighborhood in $w(\mathcal{Q}^*,\mathcal{Q})$.) Now let $\rho \geq 1$ be a constant such that $|\mu(a_i)| \leq \rho$ for all $\mu \in S$ and $i = 1,2,\dots m$, and choose $a \in A$ such that $p(a) \geq 2\rho(2n+1)$. By the Sandwich-Theorem then there is a linear functional $\mu_1 \in S$ with $\mu_1(a) \geq 2\rho(2n+1)$ as well. Now with

$$\mu = (1 - \frac{1}{2\rho})\mu_0 + \frac{1}{2\rho}\mu_1 \in S$$

we have for all $i = 1,2,\dots m$

$$|\mu(a_i) - \mu_0(a_i)| \leq \frac{1}{2\rho}(|\mu_0(a_i)| + |\mu_1(a_i)|) \leq \frac{2\rho}{2\rho} = 1,$$

whence $\mu \in U \subset S_n$. But this renders a contradiction, since

$$\mu(a) = (1 - \frac{1}{2\rho})\mu_0(a) + \frac{1}{2\rho}\mu_1(a) \geq (1 - \frac{1}{2\rho})(-n) + \frac{1}{2\rho}2\rho(2n+1) \geq (-n)+(2n+1) = n+1,$$

thus completing our proof.

3.11 Remark. The stronger statement "If $\sup\{p(a) \mid a \in A\} = +\infty$, then there is $\mu \in \mathcal{Q}^*$, $\mu \leq p$ such that $\sup\{\mu(a) \mid a \in A\} = +\infty$." does not hold in general, as we shall show in the following counterexample: Let $\mathcal{Q}$ be the vector space of finite sequences $a = (a^i)_{i \in N}$ in R (i.e. all but finitely many of the a^i are zero), endowed with the abstract neighborhood system

$$V = \{\rho > 0\}, \text{ and } (a^i) \leq (b^i) + \rho \text{ if } a^i \leq b^i + \rho \text{ for all } i \in N.$$

Then $p : \mathcal{Q} \to R$ defined by $p(a) = \sup\{a^i \mid i \in N\}$ is a u-continuous sublinear functional on $\mathcal{Q}$. Now consider the subset of $\mathcal{Q}$

$$A = \{a_n = (a_n^i) \mid n \in N\} \quad \text{such that} \quad a_n^i = \begin{cases} -n, & \text{if } i < n, \\ +n, & \text{if } i = n, \\ 0, & \text{if } i > n. \end{cases}$$

Then clearly $\sup\{p(a_n) \mid n \in N\} = +\infty$. But for every linear functional $\mu \in Q^*$ which is dominated by p we have $\mu = (\mu^i)_{i \in N}$ for non-negative numbers μ^i such that $\mu(a) = \sum_{i=1}^{\infty} \mu^i a^i$ for all $a \in Q$, and $\sum_{i=1}^{\infty} \mu^i = 1$. Thus $\mu(a_n) = -n\sum_{i=1}^{n-1} \mu^i + n\mu_n$, which tends to $-\infty$ as n tends to $+\infty$.
In particular this argument shows that $\sup\{\mu(a_n) \mid n \in N\} < +\infty$.

4. Extreme points and faces.

We are going to study now faces and extreme points in compact convex subsets of the dual cone. As usual, by (Q,V) we denote a locally convex cone, by Q^* its dual endowed with the topology $w(Q^*,Q)$. As, in general, Q^* is not embeddable in a locally convex vector space, the same holds for compact convex subsets of Q^*. So the classical results about extreme points and faces, in particular Krein-Milman's theorem, are not immediately available. We shall, however, recover some of them.

In the following, let C be a compact convex subset of Q^*. The following results will in particular apply to the polars v_Q° of abstract neighborhoods $v \in V$. We shall first recall some basic definitions.

4.1 Faces. A subset F of the compact convex set C is said to be a *face* of C, if, for $\mu \in F$, $\mu_1,\mu_2 \in C$ and $0 < \lambda < 1$

$$\mu = \lambda\mu_1 + (1-\lambda)\mu_2 \quad \text{implies} \quad \mu_1,\mu_2 \in F.$$

The intersection of any collection of faces of C is immediately seen to be a face again.

4.2 Extreme points. The element μ is called an *extreme point* of C, if $\{\mu\}$ is a singleton face in C, i.e. if for $\mu,\mu_1,\mu_2 \in C$ and $0 < \lambda < 1$

$$\mu = \lambda\mu_1 + (1-\lambda)\mu_2 \quad \text{implies} \quad \mu = \mu_1 = \mu_2.$$

We shall denote the set of extreme points of C by $Ex(C)$.

4.3 Affine functions. A function $f : C \to \overline{R}$ is said to be *affine* if for all $\mu_1,\mu_2 \in C$ and $0 < \lambda < 1$ we have

$$f(\lambda\mu_1 + (1-\lambda)\mu_2) = \lambda f(\mu_1) + (1-\lambda) f(\mu_2)$$

Continuity of f is regarded with respect to the topology $w(Q^*,Q)$ on C and the usual topology on $\overline{R}$.

4.4 Lemma. *Every continuous affine function $f : C \to \overline{R}$ attains its minimal value on some non-empty closed convex face of C.*

Proof. As the function f is continuous it attains its minimum value α on the compact set C. Thus $F = \{\mu \in C \mid f(\mu) = \alpha\}$ is a non-empty closed convex subset of C, and for

$$\mu \in F,\ \mu_1,\mu_2 \in C \text{ and } 0 < \lambda < 1, \text{ such that } \mu = \lambda\mu_1+(1-\lambda)\mu_2$$

we have $\alpha = f(\mu) = \lambda f(\mu_1)+(1-\lambda)f(\mu_2)$, whence $f(\mu_1) = f(\mu_2) = \alpha$ as well. So $\mu_1,\mu_2 \in F$, and F is indeed a face.

4.5 Lemma. *Every non-empty closed convex face F in C contains an extreme point.*

Proof. Let $\mathbf{F}$ be the collection of all non-empty closed convex faces in C which are subsets of F. We shall use Zorn's Lemma in order to show that $\mathbf{F}$ contains a minimal element: Let $\mathbf{K}$ be a linearly ordered subset of $\mathbf{F}$ and $K_0 = \bigcap_{K\in\mathbf{K}} K$. Then K_0 is not empty because of compactness, it is clearly a closed convex face, whence a lower bound for $\mathbf{K}$.
All left to show is that every minimal face F_0 in C is in fact a one-point set, hence its single element is an extreme point. Indeed, suppose we had $\mu,\nu \in F_0$ for two different elements μ and ν. Then we could find some $a \in Q$ such that $\mu(a) \neq \nu(a)$ and $\{\mu \in F_0 \mid \mu(a) = \min\{\nu(a) \mid \nu \in F_0\}\}$ would be a face smaller than F_0 contradicting its minimality.

4.6 Corollary. *Let Q be a subcone of the locally convex cone (P,V), $v \in V$. Then every $\mu \in Ex(v_Q^\circ)$ is the restriction on Q of some $\nu \in Ex(v_P^\circ)$.*

Proof. Clearly, the restriction map $\pi : P^* \to Q^*$ is is linear and continuous and maps v_P° onto v_Q° by the Extension Theorem 2.9. Thus, for $\mu \in Ex(v_Q^\circ)$ the set $F = \pi^{-1}\{\mu\} \cap v_P^\circ$ is non-empty, closed, convex, and a face of v_P° since $\nu = \lambda\nu_1+(1-\lambda)\nu_2$ for some $\nu \in F$, $\nu_1,\nu_2 \in v_P^\circ$, $0 < \lambda < 1$ implies $\mu = \pi(\nu) = \lambda\pi(\nu_1)+(1-\lambda)\pi(\nu_2)$, whence $\mu = \pi(\nu_1) = \pi(\nu_2)$, as μ is an extreme point of v_Q°. So by the preceding lemma, F contains an extreme point of v_P°.

4.7 Corollary. *Suppose that the locally convex cone (Q,V) is tightly covered by its bounded elements* (2.13). *Then the strict separation property for Q holds already with the extreme points of the polars of the respective abstract neighborhoods; more precisely: For all $a,b \in Q$ and $v \in V$ such that $a \not\leq_v b+v$, i.e $a \not\leq b+\rho v$ for some $\rho > 1$, there is a linear functional $\mu \in Ex(v_Q^\circ)$ such that $\mu(a) > \mu(b)+1$.*

Proof. By property 2.13 there is a bounded element $a' \in Q$ such that $a' \leq a$ but $a' \not\leq b+\rho v$ for some $\rho > 1$. The mapping

$$\mu \to (\mu(b)-\mu(a')) \ :\ v_Q^\circ \to \overline{\mathbb{R}}$$

is affine and continuous, and by Lemma 2.11 its minimum value is less then -1. So by the preceding lemmas there is some extreme point μ of v_Q° such that

$$\mu(b)-\mu(a') < -1, \text{ i.e. } \mu(a) \geq \mu(a') > \mu(b)+1.$$

4.8 Example. In Example 2.17 we introduced for a normed vector space $(E, \| \|)$ with unit ball B the locally convex cone Q of sets $a+\rho B$, for $a \in E$, $\rho \geq 0$. We gave a description of its dual cone. Looking for the extreme points of B°, the polar of the abstract neighborhood B in Q, they are easily identified: $Ex(B^\circ)$ consists of all functionals $\mu \oplus 1$, where μ is an extreme point of the dual unit ball of E, and of $0 \in Q^*$.

4.9 Example. A description of the dual cone of $CConv(R^n)$, the cone of all non-empty compact convex subsets of R^n, was given in Example 2.18. From the representation of $CConv(R^n)^*$ as the cone of all positive Radon measures on the dual unit sphere Y we recognize the extreme points of B°, other than 0, as the point evaluations on Y; i.e. for every $y \in Y$ there is a corresponding $\mu_y \in Ex(B^\circ)$ defined by

$$\mu_y(A) = \sup \{y(a) \mid a \in A\} \qquad \text{for all } A \in CConv(R^n).$$

5. Uniformly directed cones.

As we experienced in the previous sections a complete and handy characterization of the dual cone is available only in some special cases. In fact, even for standard examples like $Conv(R^n)$ the dual turns out to be huge and hard to handle. However, for our upcoming applications to approximation theory we shall have to deal only with certain subsets of the dual cone, in particular the extreme points of the polars of our abstract neighborhoods. In locally convex vector lattices with M-topologies it is known that those extreme points are all lattice homomorphisms. We shall recover a similar result for certain locally convex cones. We shall, however, not require our cones to contain suprema or infima but replace this by a somewhat weaker condition which is fulfilled in many of our examples. Note that all our upcoming definitions refer to the global preorder $\lesssim$.

5.1 M-uniformly directed locally convex cones. Let (Q,V) be a locally convex cone. We shall say that it is *M-uniformly down-directed* if the basic lower neighborhoods in Q are directed downward for the global preorder $\lesssim$; in detail: For each $v \in V$ and for all $a,b,c \in Q$ such that $c \lesssim a+v$ and $c \lesssim b+v$ we have $c \lesssim d+v$ for some $d \in Q$ such that $d \lesssim a$ and $d \lesssim b$.

In an analogous way we shall say that Q is *M-uniformly up-directed* if the basic upper neighborhoods in Q are directed upward for the global preorder $\lesssim$; in detail: For each $v \in V$ and for all $a,b,c \in Q$ such that $a \lesssim c+v$ and $b \lesssim c+v$ we have $d \lesssim c+v$ for some $d \in Q$ such that $a \lesssim d$ and $b \lesssim d$.

5.2 Remarks. (a) It is clear form the definition that every locally convex vector lattice with M-topology is M-uniformly up- and down-directed.

(b) If Q is a full cone, i.e. if $V \subset Q$, then Q is M-uniformly up-directed ($c+v$ is an upper bound for $v(c)$). If on the other hand $-V \subset Q$ then Q is M-uniformly down-directed ($c-v$ is a lower bound for $(c)v$).

5.3 Uniformly directed locally convex cones. Let (Q,V) be a locally convex cone. We shall say that it is *uniformly down-directed* if for each $v \in V$ there is some $v' \in V$ such that for all $a,b,c \in Q$

$c \leq a+v'$ and $c \leq b+v'$ implies $c \leq d+v$ for some $d \in Q$ such that $d \leq a$ and $d \leq b$.

In an analogous way we shall say that Q is *uniformly up-directed* if for each $v \in V$ there is some $v' \in V$ such that for all $a,b,c \in Q$

$a \leq c+v'$ and $b \leq c+v'$ implies $d \leq c+v$ for some $d \in Q$ such that $a \leq d$ and $b \leq d$.

5.4 Remarks. (a) A uniformly directed locally convex cone is directed in the usual sense with respect to the global preorder $\leq$: Let $a,b \in Q$. For $v \in V$ let $v' \in V$ be the corresponding neighborhood according to 5.3. There is some $\rho > 0$ such that $0 \leq a+\rho v'$ and $0 \leq b+\rho v'$; i.e. $0 \leq a/\rho+v'$ and $0 \leq b/\rho+v'$. If Q is uniformly down-directed we find $d \in Q$ such that $d \leq a/\rho$ and $d \leq b/\rho$; i.e. $\rho d \leq a$ and $\rho d \leq b$. If on the other hand Q is uniformly up-directed we conclude from $a/\rho \leq (a+b)/\rho+v'$ and $b/\rho \leq (a+b)/\rho+v'$ that there is some $d \in Q$ such that both $a \leq \rho d$ and $b \leq \rho d$.

(b) Definition 5.3 may be reformulated: Q is uniformly down-directed if and only if the mapping

$$(A,B) \to A \cap B \ : \ \overline{DConv(Q)} \times \overline{DConv(Q)} \to \overline{DConv(Q)}$$

is well defined (i.e. $A \cap B$ is not empty) and uniformly continuous. Indeed, suppose that this mapping is uniformly continuous and let $v \in V$. Choose $v' \in V$ such that this uniform continuity holds with the neighborhoods $(\bar{v}',\bar{v}')$ in $\overline{DConv(Q)} \times \overline{DConv(Q)}$ and $\bar{v}$ in $\overline{DConv(Q)}$. Now consider the canonical embedding

$$a \to \bar{a} = \{b \in Q \mid b \leq a\} : Q \to \overline{DConv(Q)}$$

and let $a,b,c \in Q$ as in 5.3. Then we have $(\bar{c},\bar{c}) \leq (\bar{a},\bar{b})+(\bar{v}',\bar{v}')$, whence $\bar{c} = \bar{c} \cap \bar{c} \leq \bar{a} \cap \bar{b} + \bar{v}$. in $\overline{DConv(Q)}$. So, indeed, $c \leq d+v$ for some $d \in Q$ such that $d \leq a$ and $d \leq b$. If, on the other hand, Q is uniformly down-directed, for $v \in V$ we choose $v' \in V$ as in 5.3. For $(A',B'),(A,B) \in \overline{DConv(Q)} \times \overline{DConv(Q)}$ such that $(A',B') \leq (A,B)+(\bar{v}',\bar{v}')$ let $c \in A' \cap B'$. Then $c \leq a+v'$ and $c \leq b+v'$ for some $a \in A$ and $b \in B$, whence $c \leq d+v$ for some $d \in A \cap B$ by 5.3. But this shows that $A' \cap B' \leq A \cap B + \bar{v}$, and our mapping is indeed seen to be uniformly continuous.

(c) Clearly, the uniformly directed cone Q is M-uniformly directed if the condition in 5.3 holds with $v' = v$.

5.5 Semilattices. If the locally convex cone (Q,V) contains infima (resp. suprema) with respect to the global preorder for any two of its elements they need not necessarily be uniquely determined as Q may not be separated. We shall, however, use the usual notations $a\wedge b$ and $a\vee b$ as the properties involved will always apply to all infima and suprema. In these cases we shall call Q an *$\wedge$-semilattice* or a *$\vee$-semilattice*, and a *lattice* if it is both.

The following implies in particular that every locally convex topological vector lattice is both uniformly up- and down-directed:

5.6 Proposition. *Let (Q,V) be a locally convex cone.*

(i) If Q is an $\wedge$-semilattice then it is uniformly down-directed if and only if

$$(a,b) \to a\wedge b \; : \; Q\times Q \to Q$$

is uniformly continuous with respect to the canonical quasiuniform structure.

(ii) If Q is a $\vee$-semilattice then it is uniformly up-directed if and only if

$$(a,b) \to a\vee b \; : \; Q\times Q \to Q$$

is uniformly continuous with respect to the inverse canonical quasiuniform structure.

(iii) If Q is a lattice then it is uniformly up- and down-directed if and only if the lattice operations are uniformly continuous with respect to the symmetric quasiuniform structure.

Proof. The arguments in (i) and (ii) are analogous, and (iii) is an immediate consequence of (i) and (ii). But (i) follows from 5.4 (b) if we consider the canonical embeddings

$$a \to \bar{a} \; : \; Q \to \overline{DConv(Q)}$$

and

$$(a,b) \to (\bar{a},\bar{b}) \; : \; Q\times Q \to \overline{DConv(Q)} \times \overline{DConv(Q)}$$

and the observation that for $\wedge$-semilattices we have $\bar{a} \cap \bar{b} = \overline{a\wedge b}$, .

Simple examples show that in semilattice cones we generally do not have the distributive law for the operations $\wedge$ and $\vee$

(D) $$(a+c)\wedge(b+c) = (a\wedge b)+c,$$

(D') $$(a+c)\vee(b+c) = (a\vee b)+c.$$

But this will be remedied by the following:

5.7 The semi interpolation property. Let (Q,V) be a locally convex cone. We shall say that Q has the *downward semi interpolation property* (DSIP) if for all $a,b,c,e \in Q$

$c \le a+e$ and $c \le b+e$ implies $c \le d+e$ for some $d \in Q$ such that $d \le a$ and $d \le b$.

Q has the *upward semi interpolation property* (USIP) if for all $a,b,c,e \in Q$

$c \ge a+e$ and $c \ge b+e$ implies $c \ge d+e$ for some $d \in Q$ such that $d \ge a$ and $d \ge b$.

5.8 Remarks. (a) The semi interpolation property for ordered cones is introduced in [22] and coincides with our notation of the DSIP.

(b) Every vector space trivially has the DSIP and the USIP (set $d=c\text{-}e$).

(c) If Q is an $\wedge$-semilattice (resp. a $\vee$-semilattice) then (D) (resp.(D')) holds if and only if Q has the DSIP (resp. the USIP).

The above definitions will be justified in the following section when we shall study proper linear operators between locally convex cones having those properties. We shall conclude this section looking at some of our examples in this context.

5.9 Examples. (a) $\overline{Q}\text{-}\overline{B}_Q$ is (M-)uniformly directed whenever Q is, and has the DSIP (USIP) whenever Q has. We shall show this for a uniformly down-directed cone Q (we omit the bars): Let $v \in V$ and choose v' as in (5.3).
Now let $(c\text{-}c') \le (a\text{-}a')+v'$ and $(c\text{-}c') \le (b\text{-}b')+v'$, i.e. $(c+a') \le (a+c')+v'$ and $(c+b') \le (b+c')+v'$. Adding b' resp. a' on both sides this leads to $(c+a'+b') \le (a+b'+c')+v'$ and $(c+a'+b') \le (a'+b+c')+v'$. By assumption on Q then there is some $d \in Q$ such that $d \le (a+b'+c')$ and $d \le (a'+b+c')$ and $(c+a'+b') \le d+v$. But this shows that $(c\text{-}c') \le (d\text{-}(a'+b'+c'))+v$ and both $(d\text{-}(a'+b'+c')) \le (a\text{-}a')$ and $(d\text{-}(a'+b'+c')) \le (b\text{-}b')$. So $\overline{Q}\text{-}\overline{B}_Q$ is indeed (M-)uniformly directed as well.
Now suppose that Q has the DSIP and let $(c\text{-}c') \le (a\text{-}a')+(e\text{-}e')$ and $(c\text{-}c') \le (b\text{-}b')+(e\text{-}e')$. As before we conclude $(c+a'+b'+e') \le (a+b'+c')+e$ and $(c+a'+b'+e') \le (a'+b+c')+e$. By the DSIP for Q then we find $d \in Q$ such that $(c+a'+b'+e') \le d+e$ and both $d \le (a+b'+c')$ and $d \le (a'+b+c')$. This shows that $d\text{-}(a'+b'+c') \le (a\text{-}a')$ and $d\text{-}(a'+b'+c') \le (b\text{-}b')$ as well as $(c\text{-}c') \le d\text{-}(a'+b'+c')$.

(b) Let (Q,V) be a locally convex cone and D any subcone of $\overline{DConv(Q)}$. If D contains the closed convex hull of any two of its members then it is M-uniformly up-directed and fulfills the USIP: Indeed, let $A,B,C \in D$ and $v \in V$ such that both $A \le C+\overline{v}$ and $B \le C+\overline{v}$. Then for the closed (with respect to the lower topology) convex hull D of $A \cup B$ we clearly have $A \le D$ and $B \le D$, and $D \le C+\overline{v}$ as well, as for every $d \in D$ and $v \in V$ there are elements $a \in A$ and $b \in B$ and $0 \le \lambda \le 1$ such that $d \le \lambda a+(1\text{-}\lambda)b+v$. But as $a \le c'+v$ and $b \le c''+v$ for some $c',c'' \in C$ this renders $d \le c+v$ with $c=\lambda c'+(1\text{-}\lambda)c''$. A similar straightforward argument (again take the closed convex hull of A and B for D) shows that D has the USIP as well.

(c) Reconsidering Example 2.17 we see that Q is M-uniformly up-directed as it is a full cone (Remark 5.2(b)) and has the USIP:
Let $(c+\rho_c B) \ge (a+\rho_a B)+(e+\rho_e B)$ and $(c+\rho_c B) \ge (b+\rho_b B)+(e+\rho_e B)$. This implies that $\rho_c \ge \rho_a+\rho_e$ and $\rho_c \ge \rho_a+\rho_e$. It is easily checked that our condition holds with $(c\text{-}e)+(\rho_c\text{-}\rho_e)B$.

(d) Let (Q,V) be a subcone of a cone of $\overline{R}$-valued functions as introduced in Example 2.19, i.e. the abstract neighborhoods are defined via functions $\psi_{Y,\varepsilon}$ referring to a covering Y of the domain X. Now suppose that for every set $Y \in Y$ there is a function e_Y in Q and constants $0 < \rho_Y < \sigma_Y$ such that $\rho_Y \le e_Y(y) \le \sigma_Y$ for all $y \in Y$. Then Q is uniformly up-directed: If v is the neighborhood defined by $\psi_{Y,\varepsilon}$ then the condition holds with v' defined by $\psi_{Y,\delta}$ where $\delta = (\rho_Y/\sigma_Y)\varepsilon$. Indeed, if $a \le c+v'$ and $b \le c+v'$ let $d = c+(\varepsilon/\sigma_Y)e_Y$. Then d dominates a and b on Y and $d \le c+v$. If $\rho_Y = \sigma_Y$, i.e. the functions e_Y are all constant on the respective subsets Y, we conclude $v' = v$, and Q is even M-uniformly up-directed.

A similar argument shows that Q is (M-)uniformly down-directed if it contains functions $-e_Y$ of this type.

Finally, if Q contains the pointwise infima (resp. suprema) of its elements or if it is a vector space then it has the DSIP (resp.USIP) as well (cf. Remarks 5.8(b),(c)).

5.10 Example: Continuous Q-valued functions. Let X be a topological space and (Q,V) a locally convex cone. In Ch. I, Example 2.9, we introduced the cones $C_u(X,Q)$, $C_l(X,Q)$ and $C_s(X,Q)$ of Q-valued functions on X which are continuous with respect to the upper, lower and symmetric topologies on Q. Their respective subcones of uniformly bounded below functions which we will denote by $C_{lu}(X,Q)$, $C_{ll}(X,Q)$ and $C_{ls}(X,Q)$ endowed with the neighborhoods of uniform convergence were seen to be locally convex cones. Let us reconsider them in the light of this section. Again, we investigate the uniformly down-directed case and the lower topology. Analogous results may be obtained for the up-directed case and the upper topology:

(a) Let Q be uniformly down-directed and an $\wedge$-semilattice. For $f,g \in C_{ll}(X,Q)$ it follows immediately from Proposition 5.6 that the function h defined by $h(x) = f(x) \wedge g(x)$ is an element of $C_{ll}(X,Q)$ as well: Indeed, both mappings

$$x \to (f(x),g(x)) \;:\; X \to Q\times Q \quad \text{and} \quad (a,b) \to a\wedge b \;:\; Q\times Q \to Q$$

are continuous for the respective lower topologies, and h is nothing but their composition. Thus, $C_{ll}(X,Q)$ is an $\wedge$-semilattice. The same holds for $C_{ls}(X,Q)$ because by Proposition 5.6 the mapping $(a,b) \to a\wedge b$ is even u-continuous, whence continuous with respect to the symmetric topologies on $Q\times Q$ and Q as well. So the function h is continuous for the symmetric topology on Q whenever f and g are. Moreover, as these function cones are endowed with the topology of uniform convergence on X (c.f. I.2.9), their lattice operations are uniformly continuous if and only if they are in Q. Thus, summarizing, $C_{ll}(X,Q)$ and $C_{ls}(X,Q)$ are (M-)uniformly down-directed $\wedge$-semilattices if this holds for Q. Finally, we remarked in 5.8(c) that Q has the DSIP if and only the distributive law holds for $\wedge$, and the latter clearly implies the distributive laws for our function cones. Thus, again, the DSIP holds for $C_{ll}(X,Q)$ and for $C_{ls}(X,Q)$ if and only if it holds for Q.

(b) No such statements seem to be available in general for uniformly down-directed locally convex cones which fail to be $\wedge$-semilattices.

(c) If (Q,V) is a full cone, i.e. $V \subset Q$, then clearly $C_{lu}(X,Q)$, $C_{ll}(X,Q)$ and $C_{ls}(X,Q)$ all are full cones as well, whence M-uniformly up-directed by 5.2. They are M-uniformly down-directed if $-V \subset Q$.

6. Directional operators.

We shall introduce now a class of u-continuous linear operators between uniformly directed locally convex cones which will play the part of lattice homomorphisms between vector lattices and carry many of their well-known properties.

6.1 Directional linear operators. Let (Q,V) and (P,W) locally convex cones, $T: Q \to P$ a u-continuous linear operator.
We shall say that T is *down-directional* if for all $a,b \in Q$, $x \in P$, and $w \in W$ such that $x \leq T(a)$ and $x \leq T(b)$ there is some $c \in Q$ such that $c \leq a$ and $c \leq b$ and $x \leq T(c)+w$.
In an analogous way we shall say that T is *up-directional* if for all $a,b \in Q$, $x \in P$, and $w \in W$ such that $T(a) \leq x$ and $T(b) \leq x$ there is some $c \in Q$ such that $a \leq c$ and $b \leq c$ and $T(c) \leq x+w$.

6.2 Remarks. (a) It follows immediately from the definition that the identity operator on a uniformly down-directed (resp. up-directed) cone is down-directional (resp. u-directional).

(b) If both Q and P are uniformly down-directed Definition 6.1 may be reformulated using the canonical extension of the operator $T: Q \to P$ into an operator $\overline{T}: \overline{DConv(Q)} \to \overline{DConv(P)}$ as introduced in Example 1.6(e). (For $A \in \overline{DConv(Q)}$ we set $\overline{T}(A) = \overline{T(A)}$.) Then T is seen to be down-directional if and only if for all $a,b \in Q$ we have

$$\overline{T}(A) \cap \overline{T}(B) = \overline{T}(A \cap B) \quad \text{for all } A,B \in \overline{DConv(Q)}.$$

Indeed, suppose that T is down-directional and let $x \in \overline{T}(A) \cap \overline{T}(B)$; i.e. for all $v \in V$ we have $x \leq T(a)+v$ and $x \leq T(b)+v$ for some $a \in A$, $b \in B$. But this shows that $x \in \overline{T}(A \cap B)$ as, following 6.1, for all $w \in W$ there is some $c \in A \cap B$ such that $x \leq T(c)+w$. If on the other hand the above equality holds, then for a,b,x as in 6.1 we set $A = \overline{a}$, $B = \overline{b}$. We have $x \in \overline{T}(A) \cap \overline{T}(B) = \overline{T}(A \cap B)$, whence for all $w \in W$ there is some $c \in A \cap B$, i.e. $c \leq a$ and $c \leq b$, such that $x \leq T(c)+w$.

(c) If both Q and P are uniformly down-directed $\wedge$-semilattices then the u-continuous linear operator $T: Q \to P$ is down-directional if and only if it preserves the infima, i.e. we have $T(a \wedge b) = T(a) \wedge T(b)$ for all $a,b \in Q$. Indeed, suppose T is down-directional and let $x = T(a) \wedge T(b)$. Then for every $w \in W$ there is $c \in Q$ such that $c \leq a \wedge b$ and $x = T(a) \wedge T(b) \leq T(c)+w \leq T(a \wedge b)+w$. This shows that $T(a) \wedge T(b) \leq T(a \wedge b)$; the converse inequality is obvious. On the other hand, if T preserves the infima, then for a,b,x as in 6.1 our condition holds with $c = a \wedge b$. An analogous statement holds for uniformly up-directed cones with suprema and up-directional operators.

(d) For linear functionals we easily check:
$\mu \in Q^*$ is up-directional (for up-directed Q) if and only if

$$\mu(a) \vee \mu(b) = \inf\{\mu(c) \mid a \leq c \text{ and } b \leq c\} \quad \text{for all } a,b \in Q.$$

$\mu \in Q^*$ is down-directional (for down-directed Q) if and only if

(i) $\mu(a) \wedge \mu(b) = \sup\{\mu(c) \mid c \leq a$ and $c \leq b\}$ for all $a,b \in Q$ and

(ii) in case $\mu(a) = \mu(b) = +\infty$ there is some $c \in Q$ such that $c \leq a$ and $c \leq b$ and $\mu(c) = +\infty$.

The last observation for down-directional linear functionals is somehow surprising and gives rise to the following supplement to our definition:

6.3 B-directional operators. The u-continuous linear operator $T : Q \to P$ is said to be *bd-directional* (resp. *bu-directional*) if the condition in 6.1 holds for bounded elements $x \in P$ only.
Then for bd-directional linear functionals we only need property (i) in 6.2(d).

6.4 Proposition. *The composition* ST *of two down-directional u-continuous linear operators* S *and* T *is down-directional. If* S *is bd-directional and* T *is down-directional then* ST *is bd-directional.*
A similar statement holds for up-directional u-continuous linear operators.

Proof. The proof is straightforward. We give it for the down-directional case: Let (Q,V), (P,W) and (R,U) be locally convex cones, and $T : Q \to P$ and $S : P \to R$ be down-directional operators. Then for $a,b \in Q$, $r \in R$, and $u \in U$ such that $r \leq ST(a)$ and $r \leq ST(b)$ there is some $x \in P$ such that $x \leq T(a)$ and $x \leq T(b)$ and $r \leq S(x)+u/2$. Now select $w \in W$ such that for all $x,y \in P$, $x \leq y+w$ implies $S(x) \leq S(y)+u/2$. Then, now using that T is down-directional, there is some $c \in Q$ such that $c \leq a$ and $c \leq b$ and $x \leq T(c)+w$. Combining this yields $r \leq S(x)+u/2 \leq ST(c)+u$.
If S is only supposed to be bd-directional the above argument holds for bounded elements $r \in R$, and ST is seen to be bd-directional.

Recalling that the image of a linear functional under the adjoint of an operator is nothing but the composition of this functional with the operator, we have as an immediate consequence:

6.5 Corollary. *Let* $T : Q \to P$ *be a directional u-continuous linear operator with its adjoint* T^*. *Then* $T^*(\mu)$ *is directional (resp. b-directional) in* Q^*, *whenever* μ *is directional (resp. b-directional) in* P^*.

6.6 Lemma. *Let* (Q,V) *be a locally convex cone which is uniformly down-directed with DSIP. Then for every* $\mu \in Q^*$ *and* $a,b \in Q$ *there are* $\mu_1,\mu_2 \in Q^*$ *such that* $\mu = \mu_1+\mu_2$ *and*

$$\mu_1(a)+\mu_2(b) = \sup\{\mu(c) \mid c \in Q,\ c \leq a \text{ and } c \leq b\}.$$

If Q *is uniformly up-directed with USIP we have in an analogous way*

$$\mu_1(a)+\mu_2(b) = \inf\{\mu(c) \mid c \in Q,\ a \leq c \text{ and } b \leq c\}.$$

Furthermore, if $\mu \in v^\circ$ *for some* $v \in V$, *and* $v' \in V$ *is a corresponding neighborhood according to 5.3 we have*

$$\inf\{\rho > 0 \mid \mu_1 \in \rho v'^\circ\} + \inf\{\rho > 0 \mid \mu_2 \in \rho v'^\circ\} \le 1.$$

Proof. We give the proof for the down-directed case: On $Q\times Q$ we introduce the abstract neighborhoods $\overline{v}$ $(v \in V)$ by

$$(c,d) \preceq (c',d')+\overline{v} \text{ if and only if } c \preceq c'+v \text{ and } d \preceq d'+v.$$

With the elements a and b from the statement of our lemma we consider the subcone

$$\Lambda = \{\lambda(a,b)+(c,c) \mid \lambda \ge 0 \text{ and } c \in Q\}.$$

On Λ we define the functional ϕ by $\phi(c,d) = \sup\{\mu(e) \mid e \in Q, e \preceq c \text{ and } e \preceq d\}$. Obviously we have $\phi(c+e,d+e) \ge \phi(c,d)+\phi(e,e)$ for all $e \in Q$. But the converse inequality follows immediately with the DSIP. So ϕ is in fact linear on Λ. We shall show that it is u-continuous as well with respect to the locally convex topology induced by $Q\times Q$:
Let $\mu \in v^\circ$ and v' be the neighborhood corresponding to v as in 5.3. Let $(c,d) \preceq (c',d')+\overline{v'}$, i.e. $c \preceq c'+v'$ and $d \preceq d'+v'$, and let $e \preceq c$ and $e \preceq d$ for some $e \in Q$. Then $e \preceq c'+v'$ and $e \preceq d'+v'$. By assumption on Q there is some $f \in Q$ such that $f \preceq c'$ and $f \preceq d'$ and $e \preceq f+v$. But his shows that $\mu(e) \le \mu(f)+1$, whence $\phi(c,d) \le \phi(c',d')+1$, and $\phi \in \overline{v'}^\circ_\Lambda$.
By the Extension Theorem 2.9 there is a linear functional $\overline{\phi} \in \overline{v'}^\circ_{Q\times Q}$ which extends ϕ. Now let μ_1 and μ_2 be the linear functionals on Q defined by $\mu_1(c) = \overline{\phi}(c,0)$ and $\mu_2(c) = \overline{\phi}(0,c)$. Then both μ_1 and μ_2 are in v'° and for all $c \in Q$ we have $\mu_1(c)+\mu_2(c) = \phi(c,c) = \mu(c)$, and $\mu_1(a)+\mu_2(b) = \phi(a,b)$ for the elements a and b.
To prove the last part of our statement note that for a functional $\tau \in Q^*$ and $\rho > 0$ we have $\tau \in \rho v'^\circ$ if and only if $a \le b+v'$ implies $\tau(a) \le \tau(b)+\rho$ for all $a,b \in Q$. Now assume that the last statement of our lemma is wrong. Then we find elements $a_1,b_1,a_2,b_2 \in Q$ such that $a_1 \le b_1+v'$ and $a_2 \le b_2+v'$ and

$$\overline{\phi}(a_1,a_2) = \mu_1(a_1)+\mu_2(a_2) > \mu_1(b_1)+\mu_2(b_2)+1 = \overline{\phi}(b_1,b_2)+1.$$

But this contradicts $\overline{\phi} \in \overline{v'}^\circ_{Q\times Q}$ as $(a_1,a_2) \le (b_1,b_2)+\overline{v'}$.

Now we are ready to prove the main result of this section:

6.7 Theorem. *Let* (Q,V) *be a locally convex cone which is M-uniformly down-directed with* DSIP, $v \in V$ *an abstract neighborhood. Then every extreme point* μ *of* v° *is bd-directional, i.e.*

$$\mu(a)\wedge\mu(b) = \sup\{\mu(c) \mid c \in Q,\ c \preceq a \text{ and } c \preceq b\} \quad \text{for all } a,b \in Q.$$

If Q *is M-uniformly up-directed with* USIP *then* μ *is up-directional, i.e.*

$$\mu(a)\vee\mu(b) = \inf\{\mu(c) \mid c \in Q,\ a \preceq c \text{ and } b \preceq c\} \quad \text{for all } a,b \in Q.$$

Proof. Again, we prove the down-directed case: Given $a,b \in Q$ by the preceding lemma we find $\mu_1,\mu_2 \in Q^*$ such that $\mu = \mu_1+\mu_2$ and $\mu_1(a)+\mu_2(b) = \sup\{\mu(c) \mid c \in Q,\ c \preceq a$ and

$c \leq b\}$. Moreover, as the last part of Lemma 6.6 applies with $v' = v$, we know that $\lambda_1+\lambda_2 \leq 1$ for $\lambda_1 = \inf\{\rho > 0 \mid \mu_1 \in \rho v^\circ\}$ and $\lambda_2 = \inf\{\rho > 0 \mid \mu_2 \in \rho v^\circ\}$.
If $\lambda_1 = 0$, then $c \leq d+v$ always implies $\mu_1(c) \leq \mu_1(d)$, and as $0 \leq c+\rho v$ for all $c \in Q$, with some $\rho \geq 0$, this shows $\mu_1(c) \geq 0$ as well. Furthermore, the above property shows that $\mu+\mu_1$ is an element of v° as well in this case, and we have

$$\mu = \tfrac{1}{2}(\mu+\mu_1)+\tfrac{1}{2}\mu_2, \quad \text{whence } \mu = \mu_1+\mu = \mu_2, \text{ as } \mu \text{ is extreme.}$$

But as $\mu_1(a) \geq 0$, in particular, this shows

$$\sup\{\mu(c) \mid c \in Q,\ c \leq a \text{ and } c \leq b\} = \mu_1(a)+\mu_2(b) \geq \mu_2(b) = \mu(b) \geq \mu(a) \wedge \mu(b).$$

The converse inequality is obvious.
If $\lambda_2 = 0$ we repeat the preceding argument for μ_2 in place of μ_1.
So we may assume that $\lambda_1, \lambda_2 > 0$. This implies $\mu_1 \in \lambda_1 v^\circ$ and $\mu_2 \in \lambda_2 v^\circ$ (note that this does not hold for $\lambda_1, \lambda_2 = 0$); whence $\mu_1/\lambda_1 \in v^\circ$ and $\mu_2/\lambda_2 \in v^\circ$, and as $\lambda_2 \leq 1-\lambda_1$ this shows $\mu_2/(1-\lambda_1) \in v^\circ$ as well. Thus, we have

$$\mu = \lambda_1 \frac{\mu_1}{\lambda_1} + (1-\lambda_1)\frac{\mu_2}{1-\lambda_1}, \quad \text{whence } \mu = \frac{\mu_1}{\lambda_1} = \frac{\mu_2}{1-\lambda_1}, \text{ since } \mu \text{ is extreme.}$$

But this shows that

$$\sup\{\mu(c) \mid c \in Q,\ c \leq a \text{ and } c \leq b\} = \mu_1(a)+\mu_2(b) = \lambda_1\mu(a)+(1-\lambda_1)\mu(b) \geq \mu(a) \wedge \mu(b),$$

which completes our proof.

6.8 Remark. It can be seen from examples that under the assumptions of this theorem extreme points need not be down-directional, thus the "b" may not be omitted in the downward case. If we we associate with each functional $\mu \in Q^*$ the functional $0_\mu \in Q^*$ whose value is 0, where μ is finite, and $+\infty$ where μ is, we easily check that a bd-directional μ is even down-directional if and only its associate 0_μ is bd-directional as well.

Combining Theorem 6.7 and Corollary 6.5 yields:

6.9 Corollary. *Let (Q,V) be a uniformly, and (P,W) an M-uniformly directed locally convex cone, $T : Q \to P$ a directional operator. Then $T^*(\mu)$ is (b-)directional in Q^* for every $w \in W$ and every extreme point μ of w°.*

6.10 Examples. (a) We omit the straightforward proof of the following: If $T : Q \to P$ is a (b-)directional operator, then the same holds for its canonical extension $\overline{T} : \overline{Q\text{-}B_Q} \to \overline{P\text{-}B_P}$

(b) Let (Q,V) and (P,W) be locally convex cones and $T : Q \to P$ a u-continuous linear operator with its extension $\overline{T} : \overline{DConv(Q)} \to \overline{DConv(P)}$ as in 1.6(e). Let D be a subcone of $\overline{DConv(Q)}$ which contains the closed convex hull of any two of its members. From Example 5.9(b) we know that both D and $\overline{DConv(P)}$ are M-uniformly up-directed possessing the USIP. Then $\overline{T} : D \to \overline{DConv(P)}$ is up-directional, as it preserves the suprema:

For $A,B \in D$ we have $A \vee B = \overline{\text{conv}}(A \cup B)$, the closed convex hull of $A \cup B$. Thus

$$\overline{T}(A \vee B) = \overline{T}(\overline{\text{conv}}(A \cup B)) = \overline{T(\text{conv}(A \cup B))} = \overline{\text{conv}}(T(A) \cup T(B)) = \overline{T}(A) \vee \overline{T}(B).$$

(c) Let D be as in the previous example but assume in addition that D contains the canonical embedding $\overline{Q} = \{\overline{a} \mid a \in Q\}$ of Q. Let $A \in D$ be an element which is precompact as a subset of Q with respect to the upper topology, i.e. for each $v \in V$ there are finitely many elements $a_1, a_2, \ldots a_n \in A$ such that $A \subset \bigcup_{i=1}^{n} v(a_i)$. Now, for $v \in V$ let $\overline{v}$ be the respective neighborhood for D and let μ be an extreme point of $\overline{v}^{\circ}$. By Theorem 6.7 μ is u-directional. By μ_Q we denote the restriction of μ on $\overline{Q}$. Clearly $\mu_Q \in Q^*$ (via the embedding of Q), and for the precompact set A we infer that $\mu(A) = \sup\{\mu_Q(a) \mid a \in A\}$. Indeed, let $\varepsilon > 0$ and choose $a_1, a_2, \ldots a_n \in A$ such that the neighborhoods $(\varepsilon v)(a_i)$ cover A. If B denotes the closed convex hull of the elements a_i, then by assumption $B \in D$ as well, and $\mu(B) = \sup\{\mu(a_i) \mid i=1,\ldots,n\}$ as μ is up-directional. But because $A \le B + \overline{\varepsilon v}$ in D, this shows $\mu(A) \le \mu(B) + \varepsilon$ and proves our claim. So for the extreme points of polars of neighborhoods their action on precompact subsets may be described via the dual cone of Q alone. This will turn out to be helpful, as Q^* may be well-known and handy, when D^* is not. In particular, this observation recovers and generalizes our previous investigations of the dual of $CConv(\mathbb{R}^n)$ (Examples 2.18 and 4.9) without the need for the special representation we used at the time.

(d) Let $(E, \|\ \|)$ and $(F, \|\ \|)$ be normed spaces with unit balls B and U and $Q = \{a + \rho B \mid a \in E, \rho \ge 0\}$ and $P = \{x + \lambda U \mid x \in F, \lambda \ge 0\}$ the respective locally convex cones as introduced in Example 2.17. Both cones were seen to be uniformly up-directed with USIP (Example 5.9(d)). Now let $T : E \to F$ be a surjective linear operator, $\alpha > 0$, such that $\|T(a)\| = \alpha\|a\|$ for all $a \in E$. We extend T to an operator $\overline{T} : Q \to P$ by $\overline{T}(a + \rho B) = T(a) + \alpha\rho U$. Then $\overline{T}$ is up-directional: Let $\overline{T}(a + \rho B) \le x + \kappa U$ and $\overline{T}(b + \lambda B) \le x + \kappa U$, i.e. $\|T(a) - x\| + \alpha\rho \le \kappa$ and $\|T(b) - x\| + \alpha\lambda \le \kappa$. Then $x = T(c)$ for some $c \in E$, as T is surjective, and $\|a - c\| + \rho = (1/\alpha)\|T(a) - x\| + \rho \le \kappa/\alpha$ and $\|b - c\| + \lambda = (1/\alpha)\|T(b) - x\| + \lambda \le \kappa/\alpha$, whence $a + \rho B \le c + (\kappa/\alpha)B$ and $b + \lambda B \le c + (\kappa/\alpha)B$. Furthermore, we recognize $\|T(c) - x\| + \alpha(\kappa/\alpha) = \kappa$, thus $\overline{T}(c + (\kappa/\alpha)B) \le x + \kappa U$, which proves our claim.

(e) Let X be a compact space and Q a linear subspace of $C(X)$, which contains all the constant functions. With the pointwise order and $V = \{\rho > 0\}$ (the positive constant functions) Q is a full locally convex cone. It is M-uniformly up- and down-directed (5.2(b)) with both USIP and DSIP (5.8(b)). If v denotes the neighborhood associated to $\rho = 1$, then the extreme points of v°, other than 0, are just the point evaluations on the Choquet boundary of Q (c.f. Alfsen's book [1]) for which Theorem 6.7 yields a well-known characterization.

(f) Let Q be a real or complex function algebra, i.e. an algebra of continuous (real- or complex-valued) functions on a compact space X which contains the constant functions. We endow Q with the abstract neighborhood system $V = \{\rho > 0\}$ and the preorder

$$f \le g + \rho \quad \text{if} \quad \text{Re} f(x) \le \text{Re}\, g(x) + \rho \quad \text{for all } x \in X.$$

Thus, (Q,V) is a locally convex cone which is both uniformly up- (and down-)directed, as it is a full cone. Note that, as in (e), u-continuous linear functionals may be identified with positive Radon measures on X, evaluated by their real parts. Let (P,W) be another such function algebra on the compact space Y. If we assume in addition that Q contains the complex conjugates of all its functions, i.e. their real and imaginary parts as well, then it is obvious that every u-continuous linear operator $T: Q \to P$ maps real valued into real valued functions and complex conjugates into complex conjugates. Furthermore, if T is an algebra homomorphism, i.e. multiplicative, and maps the one function into the one function, then it is up-directional: Indeed, let $f \in Q$, $h \in P$ such that both $T(f) \leq h$ and $T(-f) \leq h$, i.e. $|\mathrm{Re}\, Tf(y)| = |T(\mathrm{Re}\, f)(y)| \leq \mathrm{Re}\, h(y)$ for all $y \in Y$. Given $\varepsilon > 0$, by the Stone-Weierstraß theorem we find a polynomial $p(t)$ such that $|p(t)-|t|| < \varepsilon/2$ for all $-\rho \leq t \leq +\rho$, with $\|f\|, \|T(f)\| \leq \rho$. Then for $g = p(\mathrm{Re}\, f)+\varepsilon/2 \in Q$ we have $f \leq g$ and $-f \leq g$, and, as T is multiplicative, $T(g) = p(T(\mathrm{Re}\, f))+\varepsilon/2$ and $T(g)(y) = \mathrm{Re}\, T(g)(y) \leq |\mathrm{Re}\, Tf(y)|+\varepsilon \leq \mathrm{Re}\, h(y)+\varepsilon$ for all $y \in Y$, whence $T(g) \leq h+\varepsilon$. Finally, if a and b are any two functions in Q such that both $T(a) \leq x$ and $T(b) \leq x$ for some $x \in P$, then with $f = 1/2(a-b)$ and $h = x-T(1/2(a+b))$ we have both $T(f) \leq h$ and $T(-f) \leq h$ and repeat the above argument. The function $c = g+1/2(a+b)$ then has the desired property as in Definition 6.1. A similar argument shows that T is down-directional as well.

(g) The following shows that the SIP is indeed essential for Theorem 6.7 and may not be omitted: Let Q be the cone of all continuous convex functions on the interval $[0,1]$ with the abstract neighborhoods $V = \{\rho > 0\}$ as in (e). From 5.2(b) we know that Q is both M-uniformly up- and down-directed. It is easily seen to have the USIP but not the DSIP. For the neighborhood corresponding to $\rho = 1$ the extreme points of its polar are immediately recognized as the point-evaluations in $[0,1]$. They are indeed all up-directional but not (except those in 0 and 1) bd-directional.

Chapter III: Subcones

In this chapter we present the first of our key results, the Sup-Inf-Theorem 1.3. Throughout this chapter we assume that $\mathcal{Q}_0$ is a subcone of the locally convex cone $(\mathcal{Q},\mathcal{V})$. For an element $a \in \mathcal{Q}$ we shall define and investigate the notions of being super- resp. subharmonic with respect to this subcone (Section 1). These notions are abstract versions of the classical concepts of super- and subharmonic functions in potential theory in the vein of H. Bauer's abstract treatment of the Dirichlet problem [8]. They are one essential ingredient in Korovkin type approximation theory where they have been used extensively (c.f. [19], [20], [49], etc.). The key result is the characterization of super- and subharmonicity in the Sup-Inf-Theorem 1.3, a result that has many predecessors.

In the remaining two sections (which can be omitted at a first reading) we investigate the structure of the set S_a of all linear functionals μ in which a given element $a \in \mathcal{Q}$ is $\mathcal{Q}_0$-superharmonic (or $\mathcal{Q}_0$-subharmonic). In general, S_a need not be closed for addition nor need it be topologically closed. For this we need $\mathcal{Q}_0$ to be an (M-)uniformly directed subcone of $\mathcal{Q}$, a notion that we introduce in Section 2 and which has been prepared in the last two sections of Chapter II. The main results are 3.1 and 3.4. They easily imply the classical result of Choquet and Deny [16] on inf-stable subcones $C(X)$. The relation of this result to Korovkin type theorems had already been noticed by Bauer (see [10] and the note by Lembcke [31]).

1. Superharmonic and subharmonic elements. The Sup-Inf-Theorem.

1.1 Super- and subharmonic elements. Let $\mathcal{Q}_0$ be a subcone of the locally convex cone $(\mathcal{Q},\mathcal{V})$. Let $\mu \in \mathcal{Q}^*$ and $a \in \mathcal{Q}$. We shall say that the element a is *$\mathcal{Q}_0$-superharmonic in* μ if firstly $\mu(a)$ is finite and if secondly, for all $\nu \in \mathcal{Q}^*$,

$$\nu(b) \le \mu(b) \text{ for all } b \in \mathcal{Q}_0 \text{ implies } \nu(a) \le \mu(a).$$

In an analogous way a is said to be *$\mathcal{Q}_0$-subharmonic in* μ if $\mu(a)$ is finite and if for all $\nu \in \mathcal{Q}^*$,

$$\mu(b) \le \nu(b) \text{ for all } b \in \mathcal{Q}_0 \text{ implies } \mu(a) \le \nu(a).$$

1.2 Remarks. (a) If $a \in \mathcal{Q}$ is $\mathcal{Q}_0$-superharmonic in $\mu+\nu \in \mathcal{Q}^*$ then a has the same property in μ and in ν. Indeed, let $\phi \in \mathcal{Q}^*$ be such that $\phi(b) \le \mu(b)$ for all $b \in \mathcal{Q}_0$. Then we have $(\phi+\nu)(b) \le (\mu+\nu)(b)$ for all $b \in \mathcal{Q}_0$, whence $(\phi+\nu)(a) \le (\mu+\nu)(a)$ by assumption, and

$\phi(a) \le \mu(a)$, as $\nu(a)$ is finite. In particular, this shows that $a \in Q$ is Q_0-superharmonic in $0 \in Q^*$ if it is in some $\nu \in Q^*$.

(b) If $a \in Q$ is Q_0-superharmonic in some $\mu \in Q^*$ then a has the same property in $\lambda\mu$ for all $\lambda \ge 0$. This is obvious for $\lambda > 0$; for $\lambda = 0$ it was shown in (a).

(c) The subset of elements of Q^* in which a given $a \in Q$ is Q_0-superharmonic, however, need not be closed with respect to addition nor with respect to the topology $w(Q^*,Q)$ defined in Ch.II, 2.1. The former may be seen in the following example:
Let $Q = C[0,1]$, $V = \{\rho > 0\}$ (the positive constant functions), and Q_0 the subcone of the continuous affine functions on $[0,1]$. Then every function $f \in Q$ is seen to be Q_0-superharmonic in the point evaluations ε_0 and ε_1, as for every $\nu \in Q^*$ $\nu(g) \le g(0)$ for all $g \in Q_0$ implies $\nu = \varepsilon_0$. But only certain functions are Q_0-superharmonic in $\mu = \varepsilon_0 + \varepsilon_1$, as for $\nu = 2\varepsilon_{1/2}$ we have $\nu(g) \le \mu(g)$ for all $g \in Q_0$, but $\nu \not\le \mu$.

(d) If both a and c are Q_0-superharmonic in μ, then $a+c$ and λa $(\lambda \ge 0)$ obviously have the same property. Thus, the subset A_μ of those element of Q which are Q_0-superharmonic in μ is a subcone of Q containing Q_0 which is easily seen to be closed with respect to the symmetric topology. If, in addition, Q is a uniformly down-directed $\wedge$-semilattice in the sense of Ch. II.5, and if the functional μ is bd-directional, then A_μ is an $\wedge$-semilattice:
Let $a,c \in A_\mu$ and $\nu \in Q^*$ such that $\nu(b) \le \mu(b)$ for all $b \in Q_0$. Then $\nu(a \wedge c) \le \nu(a) \wedge \nu(c) \le \mu(a) \wedge \mu(c) = \mu(a \wedge c)$ by II.6.2(d).

In an analogous way the above remarks hold for Q_0-subharmonic elements of Q as well. The last part of (d) then applies to uniformly up-directed locally convex cones Q and to up-directional linear functionals μ.

The following is the first of our two key results:

1.3 Sup-Inf-Theorem. *Let Q_0 be a subcone of the locally convex cone (Q,V). Let $a \in Q$ and $\mu \in Q^*$ such that $\mu(a)$ is finite. Then a is Q_0-superharmonic in μ if and only if*

$$\mu(a) = \sup_{v \in V} \inf\{\mu(b) \mid b \in Q_0,\ a \preceq b+v\}.$$

In an analogous way $a \in Q$ is Q_0-subharmonic in μ if and only if

$$\mu(a) = \inf_{v \in V} \sup\{\mu(b) \mid b \in Q_0,\ b \preceq a+v\}.$$

Proof. Because some of the arguments involved differ substantially, we shall give the proofs both for the super- and the subharmonic case. We shall start with the first one:
(i) Suppose that $\mu(a) = \sup_{v \in V} \inf\{\mu(b) \mid b \in Q_0,\ a \preceq b+v\}$ and let $\phi \in Q^*$ such that $\phi(b) \le \mu(b)$ for all $b \in Q_0$. Let $\varepsilon > 0$ and choose $v \in V$ such that $\phi \in (v/\varepsilon)^\circ_Q$. Then there is some $b \in Q_0$ such that $a \preceq b+v$ and $\mu(b) \le \mu(a)+\varepsilon$, whence $\phi(a) \le \phi(b)+\varepsilon \le \mu(b)+\varepsilon \le \mu(a)+2\varepsilon$. But this shows that $\phi(a) \le \mu(a)$, as $\varepsilon > 0$ was arbitrary, and a is indeed Q_0-superharmonic in μ.
(ii) Now let μ be any functional in Q^*, $\varepsilon > 0$, and let $v \in V$ be an abstract neighborhood such that $\mu \in (v/\varepsilon)^\circ_Q$. Then for all $b \in Q_0$ such that $a \preceq b+v$ we have $\mu(a) \le \mu(b)+\varepsilon$,

whence $\inf\{\mu(b) \mid b \in \mathcal{Q}_0,\ a \preceq b+v\} > \mu(a)-\varepsilon$. But the latter shows that the condition $\mu(a) \leq \sup_{v \in V} \inf\{\mu(b) \mid b \in \mathcal{Q}_0,\ a \preceq b+v\}$ holds in any case.

(iii) Now suppose that a is $\mathcal{Q}_0$-superharmonic in μ. We have to show that this implies

$$\mu(a) \geq \sup_{v \in V} \inf\{\mu(b) \mid b \in \mathcal{Q}_0,\ a \preceq b+v\}, \qquad \text{i.e.} \qquad \mu(a) \geq \inf\{\mu(b) \mid b \in \mathcal{Q}_0,\ a \preceq b+v\}$$

for all $v \in V$. The infimum on the right hand side of the latter inequality is obviously getting larger when v gets smaller. So it suffices to consider such $v \in V$ with $\mu \in v^\circ_{\mathcal{Q}}$. Furthermore, using the Extension Theorem II.2.9, we may assume that $\mathcal{Q}$ is a full cone. Under those assumptions we define on $\mathcal{Q}$ the functional p by

$$p(c) = \inf\{\mu(b)+2\rho \mid b \in \mathcal{Q}_0,\ \rho \geq 0,\ c \preceq b+\rho v\} \text{ for all } c \in \mathcal{Q}.$$

Sublinearity is easily checked for p. It is clearly monotone as well and, because of $p(v) \leq 2$, even u-continuous. The latter property implies that $p(c) > -\infty$ for all $c \in \mathcal{Q}$, and p is indeed a sublinear $\overline{\mathbb{R}}$-valued functional in the sense of Ch II, 2.5. By the Sandwich Theorem II.2.8 there is a linear functional $\phi \in \mathcal{Q}^*$ such that

$$\phi(c) \leq p(c) \text{ for all } c \in \mathcal{Q} \text{ and } \phi(a) = p(a).$$

For all $b \in \mathcal{Q}_0$ this implies $\phi(b) \leq p(b) \leq \mu(b)$, whence $p(a) = \phi(a) \leq \mu(a)$ as well because a is $\mathcal{Q}_0$-superharmonic in μ. Thus, using the definition of p, given $0 < \varepsilon \leq 1$, we find some $b \in \mathcal{Q}_0$ and $\rho \geq 0$ such that $a \preceq b+\rho v$ and $\mu(b)+2\rho \leq \mu(a)+\varepsilon$. The former implies $\mu(a) \leq \mu(b)+\rho$ as we assumed that $\mu \in v^\circ_{\mathcal{Q}}$; the latter shows that $\mu(b)$ is finite, as $\mu(a)$ is. Combining this we get

$$\mu(b)+2\rho \leq \mu(b)+\rho+\varepsilon, \text{ whence } \rho \leq \varepsilon \leq 1.$$

But this shows that $a \preceq b+v$ and $\mu(b) \leq \mu(b)+2\rho \leq \mu(a)+\varepsilon$, and indeed

$$\mu(a) \geq \inf\{\mu(b) \mid b \in \mathcal{Q}_0,\ a \preceq b+v\}.$$

This completes our proof for the $\mathcal{Q}_0$-superharmonic case.

In the $\mathcal{Q}_0$-subharmonic case similar arguments to those in (i) and (ii) above show that a is $\mathcal{Q}_0$-subharmonic in μ, provided that $\mu(a) = \inf_{v \in V} \sup\{\mu(b) \mid b \in \mathcal{Q}_0,\ b \preceq a+v\}$, and that we have $\mu(a) \geq \inf_{v \in V} \sup\{\mu(b) \mid b \in \mathcal{Q}_0,\ b \preceq a+v\}$ in any case.

(iv) Now suppose that a is $\mathcal{Q}_0$-subharmonic in μ. We shall show that this implies $\mu(a) \leq \sup\{\mu(b) \mid b \in \mathcal{Q}_0,\ b \preceq a+v\}$ for all $v \in V$. As before we may restrict ourselves to such $v \in V$ with $\mu \in v^\circ_{\mathcal{Q}}$. Again, we may assume that $\mathcal{Q}$ is a full cone. Now we define on $\mathcal{Q}$ the functional q by

$$q(c) = \sup\{\mu(b)-2\rho \mid b \in \mathcal{Q}_0,\ \rho \geq 0,\ b \preceq c+\rho v\} \text{ for all } c \in \mathcal{Q}.$$

Superlinearity is easily checked for q, and as $0 \preceq c+\rho v$ for all $c \in \mathcal{Q}$ with some $\rho > 0$, this shows that $q(c) > -\infty$. In addition, we define the functional p on $\mathcal{Q}$ by

$$p(c) = \inf\{\lambda q(a)+2\kappa \mid \lambda, \kappa \geq 0,\ c \preceq \lambda a+\kappa v\} \text{ for all } c \in \mathcal{Q}.$$

Again, sublinearity is obvious. That all values of p are in $\overline{\mathbb{R}}$, i.e. larger than $-\infty$, is implied by the following: For $c \in \mathcal{Q}$ assume that we have $p(c) < q(c)$. Then there are $b \in \mathcal{Q}_0$, $\rho, \lambda, \kappa \geq 0$, such that $b \preceq c+\rho v$, $c \preceq \lambda a+\kappa v$ and $\lambda q(a)+2\kappa < \mu(b)-2\rho$. The former implies $b \preceq \lambda a+(\rho+\kappa)v$, whence $\lambda q(a) = q(\lambda a) \geq \mu(b)-2(\rho+\kappa)$, and $\lambda q(a)+2\kappa \geq \mu(b)-2\rho$, a clear contradiction. Thus, indeed, we have $q(c) \leq p(c)$ for all $c \in \mathcal{Q}$, and as p is easily seen to be u-continuous $(p(v) \leq 2)$, we may apply the Sandwich-Theorem II.2.8: There is $\phi \in \mathcal{Q}^*$ such that $q(c) \leq \phi(c) \leq p(c)$ for all $c \in \mathcal{Q}$, whence $q(a) = \phi(a) = p(a)$. For all $b \in \mathcal{Q}_0$ we conclude $\phi(b) \geq q(b) \geq \mu(b)$, whence $\phi(a) = q(a) \geq \mu(a)$ as well because a is $\mathcal{Q}_0$-subharmonic in μ. Now we continue as in (iii): For $0 < \varepsilon \leq 1$ we find $b \in \mathcal{Q}_0$ and $\rho \geq 0$ such that $b \preceq a+\rho v$ and $\mu(b)-2\rho \geq \mu(a)-\varepsilon$. This shows $\mu(b) \leq \mu(a)+\rho < +\infty$, and $\mu(b)-2\rho \geq \mu(b)-\rho-\varepsilon$, whence $\rho \leq \varepsilon \leq 1$. Again, this shows that $b \preceq a+v$ and $\mu(b) \geq \mu(a)-\varepsilon$, and indeed

$$\mu(a) \leq \sup\{\mu(b) \mid b \in \mathcal{Q}_0,\ b \preceq a+v\}.$$

This completes our proof.

For subcones $\mathcal{Q}_0$ of the locally convex cone $(\mathcal{Q},\mathcal{V})$ which contain the neighborhood system $\mathcal{V}$ (respectively $-\mathcal{V}$) a simpler version of the Sup-Inf-Theorem is available:

1.4 Corollary. *Let $\mathcal{Q}_0$ be a subcone of the locally convex cone $(\mathcal{Q},\mathcal{V})$ such that $\mathcal{V} \subset \mathcal{Q}_0$. Let $a \in \mathcal{Q}$ and $\mu \in \mathcal{Q}^*$ such that $\mu(a)$ is finite. Then a is $\mathcal{Q}_0$-superharmonic in μ if and only if* $\mu(a) = \inf\{\mu(b) \mid b \in \mathcal{Q}_0,\ a \preceq b\}$.

In an analogous way, if $-\mathcal{V} \subset \mathcal{Q}_0$, the element $a \in \mathcal{Q}$ is $\mathcal{Q}_0$-subharmonic in μ if and only if

$$\mu(a) = \sup\{\mu(b) \mid b \in \mathcal{Q}_0,\ b \preceq a\}.$$

Proof. We shall give the proof for the superharmonic case by demonstrating that

$$\sup_{v \in \mathcal{V}} \inf\{\mu(b) \mid b \in \mathcal{Q}_0,\ a \preceq b+v\} = \inf\{\mu(b) \mid b \in \mathcal{Q}_0,\ a \preceq b\}$$

whenever $\mathcal{Q}_0$ contains the neighborhood system $\mathcal{V}$. Clearly for every $v \in \mathcal{V}$ we have

$$\inf\{\mu(b) \mid b \in \mathcal{Q}_0,\ a \preceq b+v\} \leq \inf\{\mu(b) \mid b \in \mathcal{Q}_0,\ a \preceq b\},$$

thus

$$\sup_{v \in \mathcal{V}} \inf\{\mu(b) \mid b \in \mathcal{Q}_0,\ a \preceq b+v\} \leq \inf\{\mu(b) \mid b \in \mathcal{Q}_0,\ a \preceq b\}$$

holds in any case. To verify the converse inequality let $\varepsilon > 0$ and choose $v \in \mathcal{V}$ such that $\mu \in \varepsilon v^\circ_{\mathcal{Q}}$; i.e. $\mu(v) \leq \varepsilon$. Let $b \in \mathcal{Q}_0$ such that $a \preceq b+v$. Then by our assumption on $\mathcal{Q}_0$ the element $b' = b+v$ is contained in $\mathcal{Q}_0$ as well, and from $\mu(b') \leq \mu(b)+\varepsilon$ we infer that

$$\inf\{\mu(b) \mid b \in \mathcal{Q}_0,\ a \preceq b\} \leq \inf\{\mu(b) \mid b \in \mathcal{Q}_0,\ a \preceq b+v\}+\varepsilon,$$

whence the converse inequality holds, indeed, thus completing our proof.

Using the characterization of the closure of subsets with respect to the lower and upper topologies as given in Ch. I,3 (Proposition 3.4) we obtain as an immediate consequence of the Sup-Inf-Theorem:

1.5 Corollary. *The element $a \in \mathcal{Q}$ is $\mathcal{Q}_0$-superharmonic in $0 \in \mathcal{Q}^*$ if and only if it is contained in the closure of $\mathcal{Q}_0$ with respect to the lower topology.*

In an analogous way $a \in \mathcal{Q}$ is $\mathcal{Q}_0$-subharmonic in $0 \in \mathcal{Q}^*$ if and only if it is contained in the closure of $\mathcal{Q}_0$ with respect to the upper topology.

If all elements of the locally convex cone $(\mathcal{Q},\mathcal{V})$ are bounded, then the symmetric quasiuniform structure in the sense of Ch. I.5 defines a locally convex cone topology on $\mathcal{Q}$ as well. Let us denote this by $(Q,\overline{\mathcal{V}})$. If, as usual, Q^* denotes the dual cone of $(Q,\mathcal{V})$ recall that, following Proposition 2.21 from Chapter II, the dual of $(Q,\overline{\mathcal{V}})$ is given by Q^*-Q^*. Using this we derive:

1.6 Corollary. *If all elements of the locally convex cone $(Q,\mathcal{V})$ are bounded, then for $a \in Q$ and a subcone Q_0 of Q the following are equivalent:*

(i) *a is Q_0-superharmonic in all elements of Q^*.*

(ii) *a is Q_0-subharmonic in all elements of Q^*.*

(iii) *a is Q_0-super-(or sub-)harmonic in $0 \in Q^*$ with respect to the symmetric topology.*

(iv) *a is contained in the closure of Q_0 with respect to the symmetric topology.*

Proof. The equivalence of (i) and (ii) is obvious, as both statements mean that $\nu(b) \le \mu(b)$ for all $b \in Q_0$ implies $\nu(a) \le \mu(a)$ for all $\mu,\nu \in Q^*$. The equivalence of (iii) and (iv) is an immediate consequence of Corollary 1.5 if we apply it to the locally convex cone $(Q,\overline{\mathcal{V}})$ generated by the symmetric quasiuniform structure on Q. All left to show is the equivalence of (i) and (iii): Following Proposition II.2.21, every linear functional $\mu : Q \to \mathbf{R}$ which is u-continuous with respect to the symmetric topology may be represented as $\mu = \mu_1 - \mu_2$ with $\mu_1,\mu_2 \in Q^*$. Thus, $\mu(b) \le 0$ for all $b \in Q_0$, just means $\mu_1(b) \le \mu_2(b)$ for all $b \in Q_0$, and a is seen to be Q_0-superharmonic in 0 with respect to the symmetric topology if and only if it is Q_0-superharmonic (with respect to the original topology) in all elements of Q^*.

1.7 Examples. (a) Let $(E,\|\ \|)$ be a normed space with unit ball B and $Q = Conv(E)$ with the abstract neighborhoods as introduced in Ch. I, Example 2.8. We consider the subcone $Q_0 = \{a+\rho B \mid a \in E,\ \rho \ge 0\}$ of Q and look for Q_0-super- and subharmonic elements. Combining Corollary 1.5 with the characterization of the closure given in Ch. I.3 we see that $A \in Q$ is Q_0-superharmonic in 0 if and only if for each $\varepsilon > 0$ there is some $a+\rho B \in Q_0$ such that $A \subset a+(\rho+\varepsilon)B$, i.e. if A is bounded as a subset of E. In an analogous way A is Q_0-subharmonic in 0 if and only if for each $\varepsilon > 0$ there is some $a+\rho B \in Q_0$ such that $a+\rho B \subset A+\varepsilon B$, which of course holds for all $A \in Q$. Furthermore, as Q is M-uniformly up-directed with USIP and contains suprema (i.e. the convex hull of the union) for any two of its elements by II.5.9, we know from Theorem II.6.7 that every extreme point $\mu \in Q^*$ of the polar of the unit ball B is up-directional. Remark 1.2(d) yields that for such μ the set A_μ of Q_0-subharmonic elements is a $\vee$-semilattice and closed with respect to the symmetric topology on Q; i.e. A_μ contains the convex hull of the union of finitely many of its elements and the topological closure as a subset of E (with respect to the norm topology) of each of its member sets. Note that A_μ contains all singleton subsets of E, as Q_0 does, and that every precompact convex subset of E may be approximated (again, with respect to the norm topology) by convex sets generated by finitely many elements of E. So the above shows in particular that every precompact convex subset $A \in Q$ is Q_0-subharmonic in all such extreme points.

(b) As in (a), let $(E,\|\ \|)$ be a normed space with unit ball B and $Q = \{a+\rho B \mid a \in E, \rho \geq 0\}$ as in Ch. II, Example 2.17. Let F be a subcone of E and $Q_0 = \{b+\rho B \mid b \in F, \rho \geq 0\}$. From Ch. II, 4.8, we know that the extreme points of the polar of B consist of all functionals $\mu \oplus 1$, where μ is an extreme point of the dual unit ball of E. For another functional $\nu \oplus r \in Q^*$ then $\nu \oplus r(b+\rho B) \leq \mu \oplus 1(b+\rho B)$ for all $b \in F$, $\rho \geq 0$, means $\|\nu\| \leq 1$ and $\nu(b) \leq \mu(b)$ for all $b \in F$. So an element $a \in E$ is seen to be Q_0-superharmonic in the extreme point $\mu \oplus 1$ if for every linear functional ν on E such that $\|\nu\| \leq 1$ and $\nu(b) \leq \mu(b)$ for all $b \in F$ we have $\nu(a) \leq \mu(a)$ as well. Note that by Choquet's Theorem every such functional ν may be represented as a probability measure on the polar of B concentrated on every Baire set containig the extreme points. In the case $E = C(X)$, for some compact space X, a function $f \in E$ is Q_0-superharmonic in the point evaluation ε_x if for all Radon measures ν of norm at most 1, $\nu(g) \leq g(x)$ for all $g \in F$ implies $\nu(f) \leq f(x)$. A similar statement holds for spaces of complex valued functions,

(c) If $Q = E$ is a locally convex ordered topological vector space in its canonical representation as a locally convex cone we know from Ch. I, Corollary 3.7 that for a decreasing subcone Q_0 the closure in the lower and in the symmetric topologies (i.e. the original topology of E) coincide. Corollaries 1.5 and 1.6 then render: An element $a \in Q$ is Q_0-superharmonic in 0 if and only if it is contained in the closure of Q_0 and if and only if it is Q_0-superharmonic in all elements of Q^*. Thus, we recover some elementary facts about locally convex ordered vector spaces from our general theory.

2. Uniformly directed subcones.

We remarked in the previous section (1.2(c)) that the set of linear functionals $\mu \in Q^*$ in which a given element $a \in Q$ is Q_0-superharmonic neither needs to be closed with respect to algebraic nor topological operations. But further results about the structure of this subset may be obtained using additional assumptions on Q_0. The following definition will adapt the notion of vector sublattices to the more general situation of uniformly directed cones (compare Ch. II, sec. 5). Note that, again, all our upcoming conditions refer to the global preorder $\preceq$.

2.1 (M-)uniformly directed subcones. As before, let Q_0 be a subcone of the locally convex cone (Q,V). We shall say that Q_0 is a *uniformly down-directed subcone of* Q if for each $v \in V$ there is some $v' \in V$ such that for all $a,b \in Q_0$ and $c \in Q$ the following property holds:

If $c \preceq a+v'$ and $c \preceq b+v'$, then $c \preceq d+v$ for some $d \in Q_0$ such that $d \preceq a$ and $d \preceq b$.

In an analogous way we shall say that Q_0 is a *uniformly up-directed subcone of* Q if for each $v \in V$ there is some $v' \in V$ such that for all $a,b \in Q_0$ and $c \in Q$ the following holds:

If $a \preceq c+v'$ and $b \preceq c+v'$, then $d \preceq c+v$ for some $d \in Q_0$ such that $a \preceq d$ and $b \preceq d$.

Finally, we shall say that Q_0 is an *M-uniformly down- (resp. up-)directed subcone of Q* if the above conditions hold with $v' = v$.

2.2 Remark If Q_0 is an (M-)uniformly down-directed subcone of Q then it is (M-)uniformly down-directed in the sense of Ch. II.5. In particular, this shows that a uniformly down-directed subcone is directed downward in the usual sense with respect to the global pre-order. Furthermore, as the restriction to Q_0 of any bd-directional (c,f. Ch. II.6) linear functional $\mu \in Q^*$ is seen to be bd-directional on Q_0 as well, we have for such μ, for all $a,b \in Q_0$ $\quad \mu(a)\wedge\mu(b) = \sup\{\mu(c) \mid c \in Q_0,\ c \lesssim a \text{ and } c \lesssim b\}$.
An analogous statement holds for (M-)uniformly up-directed subcones and up-directional linear functionals.

2.3 Proposition. *Let (Q,V) be (M-)uniformly down-directed and an $\wedge$-semilattice, and let Q_0 be a subcone of Q.*

(i) *If Q_0 is inf-stable then it is an (M-)uniformly down-directed subcone of Q.*

(ii) *If Q_0 is an (M-)uniformly down-directed subcone of Q then the closure of Q_0 with respect to the symmetric topology is inf-stable.*

An analogous statement holds for subcones of (M-)uniformly up-directed $\vee$-semilattices .

Proof. We shall verify the down-directed case: (i) Suppose that Q_0 is inf-stable and let $v \in V$. According to Proposition II.5.6, we choose $v' \in V$ such the the continuity of the $\wedge$-operation in Q holds with the neighborhoods (v',v') in $Q\times Q$ and v in Q. For $a,b \in Q_0$ and $c \in Q$ as in 2.1 by assumption on Q_0 let $d = a\wedge b \in Q_0$. Then $c = c\wedge c \leq a\wedge b+v = d+v$, as desired.

(ii) Suppose that Q_0 is a uniformly down-directed subcone of Q and let a,b be in the closure of Q_0 with respect to the symmetric topology. We shall show that for every $v \in V$ the symmetric neighborhood $v(a\wedge b)\cap(a\wedge b)v$ of $a\wedge b$ meets Q_0, whence $a\wedge b$ is contained in the closure of Q_0 as well: There are elements $a',b' \in Q_0$ such that $a' \in v(a)\cap(a)v$ and $b' \in v(b)\cap(b)v$. Now we choose $v' \in V$ such that the continuity of the $\wedge$-operation in Q holds with the neighborhoods (v',v') in $Q\times Q$ and $v/2$ in Q. Thus, we have $a'\wedge b' \in (v/2)(a\wedge b)\cap(a\wedge b)(v/2)$. Following 2.1, then for $c = a'\wedge b'$ we find some $d \in Q_0$ such that $d \lesssim c \lesssim d+v/2$; i.e. $d \in (v/2)(c)\cap(c)(v/2)$. But this shows that $d \in v(a\wedge b)\cap(a\wedge b)v$.

2.4 Remarks and examples. (a) Every locally convex topological vector lattice in its canonical representation as a locally convex cone was seen to be uniformly up- and down-directed (Proposition II.5.6). Thus, the preceding result applies in this case: Every inf-stable subcone of a locally convex topological vector lattice is a uniformly down-directed subcone, and the closure (the symmetric topology coincides with the original one) of every uniformly down-directed subcone is inf-stable.

(b) Let $(E,\|\ \|)$ be a normed space with unit ball B, and let p be an L-projection on E; i.e. a linear projection such that $\|x\| = \|p(x)\|+\|x-p(x)\|$ for all $x \in E$. L-projections are well-known and investigated (c.f. Alfsen and Effros [2], Cunningham jr. [17]). Their range spaces are usually called L-summands or L-ideals. Now let $Q = \{a+\rho B \mid a \in E, \rho \geq 0\}$, the locally convex cone as introduced in Ch. II, Example 2.17. From Ch. II, 5.5(c) we know that Q is M-uniformly up-directed. Let F be the L-summand of our projection p and $Q_0 = \{b+\rho B \mid b \in F, \rho \geq 0\}$ the corresponding subcone of Q. We shall show that Q_0 is an M-uniformly up-directed subcone of Q: Indeed, let $a,b \in F$, $c \in E$ and $\rho, \rho_a, \rho_b, \rho_c \geq 0$ such that, according to 2.1, we have $(a+\rho_a B) \leq (c+\rho_c B)+\rho B$ and $(b+\rho_b B) \leq (c+\rho_c B)+\rho B$, i.e. $\|a-c\| \leq \rho+\rho_c-\rho_a$ and $\|b-c\| \leq \rho+\rho_c-\rho_b$. Now let $d = p(c) \in F$. Then we compute

$$\|a-c\| = \|p(a-c)\|+\|(a-c)-p(a-c)\| = \|p(a)-p(c)\|+\|a-p(a)+p(c)-c\| = \|a-d\|+\|d-c\|,$$

whence $\|a-d\| = \|a-c\|-\|d-c\| \leq \rho+\rho_c-\|d-c\|-\rho_a$ and $\|b-d\| \leq \rho+\rho_c-\|d-c\|-\rho_b$.

Thus, setting $\rho_d = \rho+\rho_c-\|d-c\|$ we have $a+\rho_a B \leq d+\rho_d B$ and $a+\rho_a B \leq d+\rho_d B$. Furthermore, $d+\rho_d B \leq (c+\rho_c B)+\rho B$, as $\|d-c\| = \rho_c+\rho-\rho_d$.

To give an example for the existence of such an L-projection, let $E = L^1(X,\mu)$, an integration space with its usual L^1-norm. For a μ-measurable subset Y of X let p be the multiplication of the functions in E with the characteristic function of Y. Its range F is the subspace of E of those functions vanishing on $X \backslash Y$.

(c) Let X be a compact convex set in a locally convex topological vector space and Q the cone of continuous convex functions on X provided with the abstract neighborhoods $V = \{\rho > 0\}$. Q is both M-uniformly up- and down-directed and has the USIP (c.f. Ch II, 5.2(c)). Let Q_0 be the subcone of all continuous affine functions on X. If we assume in addition that X is a Bauer simplex, i.e. a simplex such that the set of its extreme points is topologically closed (for details c.f. [1]), then it may be checked that Q_0 is even an M-uniformly down-directed subcone of Q: For every continuous convex function $c \in Q$ there is in this case a unique continuous affine function $d \in Q_0$ which dominates c and coincides with it on the extreme points of X. This is the element d which we use in 2.1.

3. Super- and subharmonicity with uniformly directed subcones.

Using the Sup-Inf-Theorem 1.3 as our main tool we shall for uniformly directed subcones Q_0 investigate the structure of subsets of linear functionals in which a given element is Q_0-super- or subharmonic. We thus generalize well-known results for locally convex topological vector lattices (c.f. [19], [20]).

3.1 Proposition. *Let Q_0 be a uniformly down-directed subcone of the locally convex cone (Q,V) and $a \in Q$. Then the set S_a of all $\mu \in Q^*$ such that a is Q_0-superharmonic in μ is either empty or a subcone and even a face of Q^*.*

An analogous statement holds with subharmonicity for uniformly up-directed subcones.

Proof. By the Remarks 1.2(a) and (b) we only have to show that $\mu+\nu \in S_a$ whenever $\mu \in S_a$ and $\nu \in S_a$. We shall use the full strength of Theorem 1.3. So, let $\mu \in S_a$ and $\nu \in S_a$. It suffices to show that

$$(\mu+\nu)(a) \geq \inf\{(\mu+\nu)(b) \mid b \in Q_0,\ a \lesssim b+v\} \text{ holds for all } v \in V.$$

Thus for $v \in V$ let $v' \in V$ be as in 2.1. Given $\varepsilon > 0$, then by Theorem 1.3, as a is Q_0-superharmonic in both μ and ν, there are elements b_1 and b_2 in Q_0 such that

$$a \lesssim b_i+v',\ i=1,2, \text{ and } \mu(b_1) \leq \mu(a)+\varepsilon/2 \text{ and } \nu(b_2) \leq \nu(a)+\varepsilon/2.$$

Now by assumption on Q_0 we find $b \in Q_0$ such that $b \lesssim b_i$, $i=1,2$, and $a \lesssim b+v$. But this shows that $(\mu+\nu)(b) = \mu(b)+\nu(b) \leq \mu(b_1)+\nu(b_2) \leq (\mu+\nu)(a)+\varepsilon$ and completes our proof.

Corresponding results for the topological closure of the subset S_a are not available in general. However, we have:

3.2 Proposition. *Let Q_0 be a uniformly down-directed subcone of the locally convex cone (Q,V). Let $a \in Q$ and $\mu \in Q^*$ such that for every neighborhood U of 0 in the $w(Q^*,Q)$-topology of Q^* there are functionals $\nu \in Q^*$ and $\tau \in U$ such that $\mu = \nu+\tau$ and a is Q_0-superharmonic in ν. Then a is Q_0-superharmonic in μ as well.*
An analogous statement holds with subharmonicity for uniformly up-directed subcones.

Proof. Let $\mu \in Q^*$ with the above assumptions. As before we shall use Theorem 1.3 in both directions: Let $v \in V$ and choose $v' \in V$ as in 2.1. As a is Q_0-superharmonic in some $\nu \in Q^*$, it is in $0 \in Q^*$, and by Theorem 1.3 there is $b_1 \in Q_0$ such that $a \lesssim b_1+v'$. Now, given $\varepsilon > 0$, there are $\nu,\tau \in Q^*$ such that $\mu = \nu+\tau$; a is Q_0-superharmonic in ν, and both $|\tau(a)|, |\tau(b_1)| < \varepsilon/3$. This shows in particular that $\mu(a)$ is finite, as $\nu(a)$ is. Now, again using Theorem 1.3, choose $b_2 \in Q_0$ such that $a \lesssim b_2+v'$ and $\nu(b_2) \leq \nu(a)+\varepsilon/3$. As Q_0 is a uniformly down-directed subcone of Q then there is some $b \in Q_0$ such that $b \lesssim b_i$, $i=1,2$, and $a \lesssim b+v$. But this shows that $\mu(b) = \nu(b)+\tau(b) \leq \nu(b_2)+\tau(b_1) \leq \nu(a)+\varepsilon/3+\varepsilon/3 \leq \mu(a)+\varepsilon$ and completes our proof. The argument in the subharmonic case is similar.

The above implies a well-known result about the classical sequences spaces l^p which may for example be found in [12], [19] and [20]:

3.3 Corollary. *Let Q be the sequence space l^p for some $1 \leq p < \infty$ with its usual, i.e. componentwise, order and its dual cone Q^*, i.e. the cone of all positive sequences in l^q $(1/p+1/q = 1)$. Let Q_0 be an inf-stable subcone of Q. If $a \in Q$ is Q_0-superharmonic in all atoms (i.e the unit vectors $(0,..,0,1,0,..)$) of l^q, then it is Q_0-superharmonic in all elements of Q^*.*

Proof: As Q is a vector lattice, any inf-stable subcone Q_0 of Q was seen to be uniformly down-directed (Remark 2.4(a)). Let $a \in Q$ be Q_0-superharmonic in all atoms of l^q. Then it is Q_0-superharmonic in all finite positive sequences by Proposition 3.1. Furthermore, as it is easily checked that every positive sequence in l^q may be approximated in the sense of Proposition 3.2 by finite positive sequences, a is seen to be Q_0-superharmonic in all elements of Q^*.

In the following we shall investigate $w(Q^*,Q)$-compact convex subsets K of Q^*. We shall establish that under certain conditions for uniformly directed subcones, for elements of Q, super- (resp. sub-)harmonicity in the extreme points of K implies this property on the whole set. As we are going to apply Choquet's Theorem which only holds for finite continuous functions (for details we globally refer to Alfsen's book [1]) we generally have to restrict to this case. The elements of Q will be considered as finite continuous affine functions on K. The Choquet boundary of the subspace of $C(K)$ spanned by Q then coincides with $Ex(K)$, the set of the extreme points of K.

3.4 Theorem. *Let Q_0 be a uniformly down-directed subcone of the locally convex cone (Q,V). Let K be a compact convex subset of Q^* such that all elements of Q are finite on K. Let $Ex(K)$ be at most countable. If $a \in Q$ is Q_0-superharmonic in all extreme points of K then it is Q_0-superharmonic in all elements of K.*
An analogous statement holds for uniformly up-directed subcones and subharmonic elements.

Proof. Again, we give the proof for the superharmonic case: Using the notations of the theorem, let μ be any element of K. It follows from Choquet's Theorem ($Ex(K)$ is an F_σ-set) that μ has an integral representation as a probability measure concentrated on the countable subset $Ex(K)$. But this implies that μ may be expressed as

$$\mu = \sum_{i=1}^{\infty} \lambda_i \nu_i\,, \quad \text{with} \quad \nu_i \in Ex(K) \quad \text{and} \quad \lambda_i \geq 0 \quad \text{such that} \quad \sum_{i=1}^{\infty} \lambda_i = 1.$$

Convergence is meant to be with respect to $w(Q^*,Q)$. Now we apply Proposition 3.2: Given any 0-neighborhood U in this topology, we find $n \in N$ such that $\tau = \sum_{i=n+1}^{\infty} \lambda_i \nu_i \in U$. The element $a \in Q$ is Q_0-superharmonic in $\nu = \sum_{i=1}^{n} \lambda_i \nu_i$ by Proposition 3.1, whence in $\mu = \nu + \tau$ by Proposition 3.2.

In fact, the assumptions on K in the above theorem might be slightly weakened: Following the preceding proof, we only need that every element of K may be represented as a convex combination of countably many extreme points. This property may hold even if $Ex(K)$ is not countable. A stronger statement, however, is available for M-uniformly directed subcones. The following is our main result of this section and generalizes similar results for locally convex vector lattices with M-topologies as in [19] and [20].

3.5 Theorem. *Let Q_0 be an M-uniformly down-directed subcone of the locally convex cone (Q,V). Let K be a compact convex subset of Q^* such that all elements of Q are finite on K. Let B be either a Baire or an F_σ-subset of K which contains $Ex(K)$. If $a \in Q$ is*

Q_0-superharmonic in all points of B then it is also Q_0-superharmonic in all elements of K. An analogous statement holds for uniformly up-directed subcones and subharmonic elements.

Proof. We shall give the proof for the superharmonic case using using Theorem 1.3 in both directions: Using the notations of the theorem, let μ be any element of K. We have to show that $\mu(a) \geq \inf\{\mu(b) \mid b \in Q_0,\ a \lesssim b+v\}$ holds for all $v \in V$.
Given $\varepsilon > 0$, as a is Q_0-superharmonic in the elements of B, by Theorem 1.3 for every $\beta \in B$ we find some $b_\beta \in Q_0$ such that $\beta(b_\beta) \leq \beta(a)+\varepsilon/2$ and $a \lesssim b_\beta+v$. Furthermore, for every finite subset $C \subset B$, using the fact that Q_0 is an M-uniformly directed subcone (condition 2.1 obviously holds for M-uniformly directed subcones with finitely many elements of Q_0 in place of a and b), we find an element $b_C \in Q_0$ such that

$$b_C \lesssim b_\beta \text{ for all } \beta \in C \text{ and } a \lesssim b_C+v.$$

Using this element b_C, we now define on K the continuous affine function f_C and the continuous function g_C by

$$f_C(\nu) = \nu(b_C) \quad\text{and}\quad g_C(\nu) = \inf\{\nu(b_\beta) \mid \beta \in C\} \quad \text{for all } \nu \in K.$$

By $\mathcal{C}$ we denote the set of all finite subsets of B, which is ordered by inclusion and directed upward. The collection of functions $\{g_C \mid C \in \mathcal{C}\}$, however, is directed downward, and as $\beta(b_\beta) \leq \beta(a)+\varepsilon/2$, we have

$$\inf\{g_C(\beta) \mid C \in \mathcal{C}\} \leq \beta(a)+\varepsilon/2 \text{ for all } \beta \in B.$$

Now, using Choquet's Theorem for our functional $\mu \in K$, we find a probability measure Φ which is concentrated on B and represents μ, i.e $\Phi(h) = h(\mu)$ for all continuous affine functions h on K. In particular, if h_a denotes the function on K defined by $h_a(\nu) = \nu(a)$, we have $\Phi(h_a) = \mu(a)$. As the set $\{g_C \mid C \in \mathcal{C}\}$ is directed downward and as $f_C \leq g_C$ on K, using Fatou's Lemma, we conclude that $\inf\{\Phi(f_C) \mid C \in \mathcal{C}\} \leq \inf\{\Phi(g_C) \mid C \in \mathcal{C}\} \leq \Phi(\inf\{g_C \mid C \in \mathcal{C}\}) \leq \Phi(h_a)+\varepsilon/2 = \mu(a)+\varepsilon/2$. Thus, as all the functions f_C were continuous and affine, we find some $C_0 \in \mathcal{C}$ such that

$$\Phi(f_{C_0}) = f_{C_0}(\mu) = \mu(b_{C_0}) \leq \mu(a)+\varepsilon.$$

And as $b_{C_0} \in Q_0$ and $a \lesssim b_{C_0}+v$, this proves our claim.

The following application of our preceding theorem yields a well-known result due to Choquet and Deny [16]. The close connection of their theorem to Korovkin theory had been remarked first by H. Bauer [11]:

3.6 Corollary. *Let X be a compact space and let Q_0 be an inf-stable subcone of $Q = C(X)$. The function $f \in Q$ is contained in the closure (with respect to the usual sup-norm topology) of Q_0 if and only if, for all $x \in X$ and all positive Radon measures μ on X,*

$$\mu(g) \leq g(x) \text{ for all } g \in Q_0 \text{ implies } \mu(f) \leq f(x).$$

Proof. As usual, we endow Q with the pointwise order and the abstract neighborhoods $V = \{\rho > 0\}$. Then Q^* coincides with the cone of all positive Radon measures on X, and any inf-stable subcone Q_0 of Q was seen to be an M-uniformly down-directed subcone (Remark 2.4(a)). If the function $f \in Q$ is contained in the closure of Q_0 then the condition of the corollary clearly holds. On the other hand, this conditions means that f is Q_0-superharmonic in all point-evaluations in X, i.e. in the extreme points of the compact convex set of probability measures on X. Theorem 3.5 yields that every such f is also Q_0-superharmonic in all probability measures on X, whence in all of Q^*. Finally, Corollary 1.6 shows that f is contained in the closure of Q_0 with respect to the symmetric (i.e. the usual sup-norm) topology on Q.

A similar statement is possible for inf-stable subcones Q_0 of l^p-spaces using Corollaries 3.3 and 1.6. This result may for example be found in [20]:

3.7 Corollary: *Let Q_0 be an inf-stable subcone of the sequence space l^p, for $1 \le p < \infty$. The sequence $a = (a_i) \in l^p$ is contained in the closure of Q_0 if and only if for all $(x_i) \in l^q$ $(1/p + 1/q = 1)$ and all $n \in \mathbb{N}$*

$$\sum_{i=1}^{\infty} b_i x_i \le b_n \quad \text{for all } (b_i) \in Q_0 \quad \text{implies} \quad \sum_{i=1}^{\infty} a_i x_i \le a_n .$$

3.8 Example. For a Bauer simplex X and the cone of continuous convex functions Q on X we showed in Example 2.4(c) that the subcone Q_0 of continuous affine functions is M-uniformly down-directed. It is easily checked that every function f in Q is Q_0-superharmonic in all point evaluations in the extreme points of X. As $Ex(X)$ is compact in this case, it follows from Theorem 3.5 that f is even Q_0-superharmonic in all probability measures on $Ex(X)$. For readers familiar with Choquet theory this may well be understood classically: Let μ be a probability measure on $Ex(X)$ and let $\nu \in Q^*$, i.e. a positive Radon measure ν on X, such that $\nu(h) \le \mu(h)$ for all $h \in Q_0$. This means that μ and ν have the same barycentre, i.e. coincide with the same point evaluation on the affine functions. As μ is concentrated on the extreme points, it is a maximal measure with respect to the Choquet ordering. But as X is a Bauer simplex, those maximal measures are unique, and μ dominates ν in this ordering. But this means $\nu(f) \le \mu(f)$ for all continuous convex functions on X.

3.9 Example $Q = Conv(E)$**.** Let E be a locally convex topological vector space with 0-neighborhood base V. Then $Q = Conv(E)$ is a locally convex cone in the sense of Ch. I,2.8, ordered by inclusion. It is M-uniformly up-directed with USIP (Ch. II,5.5(b)). Let $Q_0 = E$, i.e. the subcone of singleton elements and Q_1 be any a subcone of precompact convex subsets of E which contains the convex hull, i.e.the supremum, of any two of its members. Then Q_1 is an M-uniformly up-directed subcone of Q. We shall list some observations about Q_0- and Q_1-super- and subharmonicity in this case. Recall that for every $\mu \in Q^*$ which is up-directional and its restriction μ_E to E (i.e. to Q_0) we have $\mu(A) = \sup\{\mu_E(a) \mid a \in A\}$ for precompact subsets A of E (Ch. II,6.10(a)).

(i) The set $A \in Q$ is Q_0-superharmonic in $0 \in Q^*$ if and only if it is contained in the closure of Q_0 with respect to the lower topology, which coincides with Q_0 if E is a Hausdorff space.
(ii) It follows from Theorem 1.3 that every non-empty convex subset of E is Q_0-subharmonic in 0. If $\mu \in Q^*$ is an extreme point in the polar of some neighborhood then it is up-directional, and the set of elements of Q which are Q_0-subharmonic in μ is sup-stable (Remark 1.2(d)). i.e. contains all precompact convex subsets of E.
(iii) If Q_1 contains Q_0 as a subcone then it follows from Theorem 1.3 that the elements of Q which are Q_1-superharmonic in 0 are precisely the precompact convex subsets of E, and every non-empty convex subset of E is Q_1-subharmonic in 0.
(iv) Now let $\mu \in Q^*$ be an extreme point in the polar of some neighborhood. The convex set $A \in Q$ is Q_1-subharmonic in μ if for every $\nu \in Q^*$ such that $\mu(B) = \sup\{\mu_E(b) \mid b \in B\} \leq \nu(B)$ for all $B \in Q_1$ we have $\mu(A) \leq \nu(A)$ as well. If A is bounded and Q_1-subharmonic in all such extreme points, then by Theorem 3.5 it is Q_1-subharmonic in all elements of Q^*, whence contained in the closure of Q_1 with respect to the symmetric topology by Corollary 1.6. If, in particular, Q_1 contains Q_0 as a subcone and if μ is up-directional, then $\mu(B) \leq \nu(B)$ for all $B \in Q_1$ means $\mu_E = \nu_E$, as $\sup\{\nu_E(b) \mid b \in B\} \leq \nu(B)$, and we have $\mu(A) \leq \nu(A)$ for every precompact convex subset of E. This recovers that the closure of Q_1 with respect to the symmetric topology in this case is nothing but the subcone of the precompact members of Q.

3.10 Example. A result similar to Corollaries 3.6 and 3.7 is possible for L-summands in normed spaces as introduced in Example 2.4 (b): Let $(E, \|\ \|)$ be a normed space with unit ball B and dual space E', and let F be an L-summand in E. For the suitable locally convex cones Q and Q_0 as introduced in Example 2.4 Q_0 was seen to be an M-uniformly up-directed subcone of Q. If we assume in addition that the extreme points of the polar of B form a Baire or an F_σ-subset with respect to the usual weak*-topology of E' then we may immediately apply Theorem 3.5 and Corollary 1.6. Recall (c.f. Example II,4.8) that the extreme points of the polar of B in Q^* were just the functional $\mu \oplus 1$, where μ is an extreme point of the polar of B in E':
The element $e \in E$ is contained in the closure of F with respect to the norm topology if and only if for all $\mu, \nu \in E'$ such that μ is an extreme point of the dual unit ball

$$\nu(f) = \mu(f) \text{ for all } f \in F \text{ implies } \nu(e) = \mu(e).$$

Chapter IV: Approximation

In the first section of this chapter we present the basic convergence result of this whole work. Its essential flavor is the following: Let $\mathcal{Q}_0$ be a set of *test elements* in a locally convex cone $(\mathcal{Q},\mathcal{V})$. If an element $a \in \mathcal{Q}$ is $\mathcal{Q}_0$-superharmonic with respect to a suitably large family K of u-continuous linear functionals on $\mathcal{Q}$, then

$$T_\alpha(b) \uparrow b \quad \text{for all } b \in \mathcal{Q}_0 \text{ implies } T_\alpha(a) \uparrow a,$$

whenever $(T_\alpha)_{\alpha \in A}$ is a net of equicontinuous linear operators on $\mathcal{Q}$. The convergence $\uparrow$ is understood with respect to the upper topology. This situation is clearly modelled after Korovkin's classical approximation theorem, but we do not impose any additional requirements (positivity, contractivity, etc.) on the operators under consideration.

Thus, $\mathcal{Q}_0$-superharmonicity is a criterion for $\mathcal{Q}_0$ to be a test set for the convergence for the convergence of $T_\alpha(a)$ towards a for any net of equicontinuous operators. We shall always assume that $\mathcal{Q}_0$ is a subcone of $\mathcal{Q}$, but it is clear that the above conditions need to be checked only for a subset generating $\mathcal{Q}_0$.

Together with our results on superharmonicity in Chapter III we can derive a large number of very different Korovkin type theorems from a simplified version of our general result, as we shall see in Section 2.

The subsequent Chapters V and VI are devoted to cone-valued functions which generate locally convex cones in various ways. Most of our Korovkin type theorems in these situations are new. Chapter VI, in particular, investigates quantitative results using the full strength of the General Convergence Theorem of this chapter.

Throughout this chapter we are going to use the following notations:

$(\mathcal{Q},\mathcal{V})$ and $(\mathcal{P},\mathcal{W})$ are locally convex cones,

$\mathcal{Q}^*$ and $\mathcal{P}^*$ their dual cones respectively,

$\mathcal{Q}_0$ is a subcone of test elements in $\mathcal{Q}$.

1. The Convergence Theorem.

For the statement of our general results we need the appropriate notion of equicontinuity and, for technical reasons, a generalization of the notion of superharmonicity in Chapter III.1.

1.1 Equicontinuity. A set E of linear operators $T: Q \to P$ is called *uniformly equicontinuous (u-equicontinuous, for short)*, if for every $w \in W$ there is a $v \in V$ such that for all $a,b \in Q$ and $T \in E$

$$a \le b+v \quad \text{implies} \quad T(a) \le T(b)+w.$$

We shall apply this definition in particular to nets $(T_\alpha)_{\alpha \in A}$ of linear operators.

1.2 Remarks. (a) A subset K of Q^* (its elements are u-continuous linear operators from Q into $\overline{R}$) is u-equicontinuous if and only if there is some $v \in V$ such that for all $a,b \in Q$ and $\mu \in K$

$$a \le b+v \quad \text{implies} \quad \mu(a) \le \mu(b)+1.$$

Thus, K is u-equicontinuous if and only if it is contained in the polar v_Q° of some neighborhood $v \in V$.

(b) If E is a u-equicontinuous set of linear operators from Q into P and if $w \in W$ and $v \in V$ are as in 1.1 then we observe that $T^*(w_P^\circ) \subset v_Q^\circ$ holds for the adjoint T^* of every operator $T \in E$ (see II.2.15).

1.3 Strictly separating sets of linear functionals. Let $K \subset Q^*$ be a set of u-continuous linear functionals on Q. We shall say that *K is strictly separating relative to a given neighborhood $v \in V$*, if the following holds:

(SP') There is a u-equicontinuous subset K_0 of K and some $\rho > 0$ such that for all $a,b \in Q$

$$a \not\le b+v \quad \text{implies} \quad \mu(a) > \mu(b)+\rho \quad \text{for some} \quad \mu \in K_0.$$

If (SP') holds for every $v \in V$, then we say that *K is strictly separating for Q*.

1.4 Remarks. (a) The reader should recall that, in II.2.12, we had defined the strict separation property (SP) for a locally convex cone (Q,V) as follows:

(SP) For all $a,b \in Q$

$$a \not\le_v b+v \quad \text{implies} \quad \mu(a) > \mu(b)+1 \quad \text{for some} \quad \mu \in v_Q^\circ.$$

It is easy to see that (SP) implies (SP') for $K = Q^*$ and all $v \in V$. In II.2.14 we had shown that (SP) is guaranteed if Q is tightly covered by its bounded elements.

(b) If $K \subset Q^*$ is strictly separating relative to $v \in V$, then K is also strictly separating relative to every λv with $\lambda > 0$ and to every neighborhood $w \in V$ such that $v \le w$. Thus, for K to be strictly separating for Q it suffices to check condition (SP') for a neighborhood base in Q.

We are interested in subsets $K \subset Q^*$ as small as possible that still are strictly separating. The following is an immediate consequence of Corollary II.4.7:

1.5 Proposition. *Let Q be tightly covered by its bounded elements, and let K be a subset of Q^*. If for $v \in V$ there is some $0 < \rho \le 1$ such that $\bigcup_{\rho \le \lambda \le 1/\rho} \lambda K$ contains the extreme points of v°, then K is strictly separating relative to v.*

The converse of this statement does not hold in general. However, we have:

1.6 Proposition. *Suppose that all elements of the locally convex cone* (Q,V) *are bounded. Let* $v \in V$ *and* K *be a closed subset of* v°_{Q} *such that (SP') holds with* $\rho = 1$. *Then* K *contains the non-zero extreme points of* v°_{Q}.

Proof. As $Q\text{-}Q$ with its symmetric topology is a locally convex vector space in this case, its dual is given by $Q^*\text{-}Q^*$, and the topology $w(Q^*,Q)$ coincides with the usual weak*-topology; we may apply classical tools: By the Hahn-Banach Separation Theorem our assumption on K guarantees that the closed convex hull of $K \cup \{0\}$ comprises all of v°_{Q}, whence the compact set $K \cup \{0\}$ contains the extreme points of v°_{Q} by Krein-Milman's Theorem.

1.7 The eventual preorder for nets. Let $(a_\alpha)_{\alpha \in A}$ and $(b_\alpha)_{\alpha \in A}$ be nets in Q with the same index set A. For a fixed neighborhood $v \in V$ we shall say that (a_α) *is eventually v-below* (b_α), and we write

$$(a_\alpha) \lesssim_v (b_\alpha) \text{ if and only if for every } \rho > 0 \text{ there is some } \alpha_0 \in A \text{ such that}$$
$$a_\alpha \le b_\alpha + \rho v \quad \text{for all } \alpha \ge \alpha_0.$$

Furthermore, we say that (a_α) *is eventually below* (b_α), and we write

$$(a_\alpha) \lesssim (b_\alpha) \text{ if and only if } (a_\alpha) \lesssim_v (b_\alpha) \text{ for all } v \in V.$$

For constant nets $a_\alpha = a$ and $b_\alpha = b$ these definitions lead to the local and global preorders for the elements of Q as defined in Ch. I, 3.1.

If we only take $b_\alpha = b$ to be a constant net, then $(a_\alpha) \lesssim b$ just means that the net (a_α) converges to b with respect to the upper topology (see I.2.2). We write

$a_\alpha \uparrow b$	if	$(a_\alpha) \lesssim b$, i.e. if the net (a_α) converges to b in the upper topology.
$a_\alpha \downarrow b$	if	$b \lesssim (a_\alpha)$, i.e. if the net (a_α) converges to b in the lower topology.
$a_\alpha \to b$	if	$a_\alpha \uparrow b$ and $a_\alpha \downarrow b$, i.e. if the net (a_α) converges to b in the symmetric topology.

Now let Q_0 be a subcone of Q. In Chapter III we had defined an element $a \in Q$ to be Q_0-superharmonic in a fixed $\mu \in Q^*$, if $\mu(a) < +\infty$ and if, for all $\nu \in Q^*$

$$\nu(b) \le \mu(b) \text{ for all } b \in Q_0 \text{ implies } \nu(a) \le \mu(a).$$

For technical reasons we need a generalization of this notion:

1.8 Super- and subharmonicity on subsets of Q^*. Let Q_0 be a subcone of Q and $K \subset Q^*$. An element $a \in Q$ is called Q_0*-superharmonic on* K if, firstly, $\mu(a)$ is finite for every $\mu \in K$ and if, secondly, for all u-equicontinuous nets $(\nu_\alpha)_{\alpha \in A}$ in Q^* and $(\mu_\alpha)_{\alpha \in A}$ in K with the same index set A

$$(\nu_\alpha(b)) \lesssim (\mu_\alpha(b)) \text{ for all } b \in Q_0 \text{ implies } (\nu_\alpha(a)) \lesssim (\mu_\alpha(a)) \text{ in } \overline{R}.$$

We say that a is *sequentially $\mathcal{Q}_0$-superharmonic on K* if the above conditions holds for sequences $(A = \mathbb{N})$ only.
In an analogous way a is said to be *(sequentially) $\mathcal{Q}_0$-subharmonic on K* if for nets (sequences) $(\nu_\alpha)_{\alpha \in A}$ in $\mathcal{Q}^*$ and $(\mu_\alpha)_{\alpha \in A}$ in K as above

$$(\mu_\alpha(b)) \lesssim (\nu_\alpha(b)) \quad \text{for all } b \in \mathcal{Q}_0 \quad \text{implies} \quad (\mu_\alpha(a)) \lesssim (\nu_\alpha(a)).$$

1.9 Remarks. (a) Note that we have $(\nu_\alpha(b)) \lesssim (\mu_\alpha(b))$ for all $b \in \mathcal{Q}_0$, if for every $b \in \mathcal{Q}_0$ there is there is some $\alpha_0 \in A$ such that

$$\nu_\alpha(b) \le \mu_\alpha(b)+1 \quad \text{for all } \alpha \ge \alpha_0.$$

(b) If $a \in \mathcal{Q}$ is $\mathcal{Q}_0$-superharmonic on the subsets $K_1, K_2, \ldots, K_n$ of $\mathcal{Q}^*$, then it is also $\mathcal{Q}_0$-superharmonic on their union K: Indeed, let $(\mu_\alpha)_{\alpha \in A}$ be a net in K and $(\nu_\alpha)_{\alpha \in A}$ in $\mathcal{Q}^*$ such that $(\nu_\alpha(b)) \lesssim (\mu_\alpha(b))$ for all $b \in \mathcal{Q}_0$. If we had not $(\nu_\alpha(a)) \lesssim (\mu_\alpha(a))$, then we could find $\delta > 0$ and a subnet $(\mu_\beta)_{\beta \in B}$ of $(\mu_\alpha)_{\alpha \in A}$ contained in one of the sets K_{i_0} such that $\nu_\beta(a) \ge \mu_\beta(a)+\delta$ for all $\beta \in B$, contradicting the assumption on K_{i_0}.

We still have to justify that the old notion of super- resp. subharmonicity on points of $\mathcal{Q}^*$ as recalled above coincides with the preceding definition for one-point sets K. This is an implication of the following:

1.10 Proposition. *Let $\mathcal{Q}_0$ be a subcone of $\mathcal{Q}$. Let K be a subset of $\mathcal{Q}^*$ such that the intersection of K with the polar of each neighborhood is $w(\mathcal{Q}^*,\mathcal{Q})$-closed. Then the element $a \in \mathcal{Q}$ is $\mathcal{Q}_0$-super- (resp. sub-) harmonic on K if and only if it is $\mathcal{Q}_0$-super- (resp. sub-) harmonic in all elements of K.*

Proof. Clearly, both properties require a to be finite in all elements of K. Furthermore, as $\mathcal{Q}_0$-superharmonicity on the set K obviously requires $\mathcal{Q}_0$-superharmonicity in all elements of K (choose constant nets), all left to show is that the latter property implies the former as well: Let $(\nu_\alpha)_{\alpha \in A}$ and $(\mu_\alpha)_{\alpha \in A}$ be nets as in 1.7, i.e. $(\nu_\alpha(b)) \lesssim (\mu_\alpha(b))$ for all $b \in \mathcal{Q}_0$. If $(\nu_\alpha(a)) \lesssim (\mu_\alpha(a))$ would not hold, then we could find some $\delta > 0$ and subnets $(\nu_\beta)_{\beta \in B}$ and $(\mu_\beta)_{\beta \in B}$ such that $\nu_\beta(a) \ge \mu_\beta(a)+\delta$ for all $\beta \in B$. Furthermore, because both nets are u-equicontinuous, i.e. contained in the $w(\mathcal{Q}^*,\mathcal{Q})$-compact polar of some neighborhood of $\mathcal{Q}$, and because the intersection of K with this polar was supposed to be $w(\mathcal{Q}^*,\mathcal{Q})$-closed, we may assume in addition that both subnets $(\nu_\beta)_{\beta \in B}$ and $(\mu_\beta)_{\beta \in B}$ are convergent, i.e.

$$\nu_\beta \to \nu \quad \text{and} \quad \mu_\beta \to \mu \quad \text{for some } \nu \in \mathcal{Q}^* \text{ and } \mu \in K.$$

But this implies $\nu(b) \le \mu(b)$ for all $b \in \mathcal{Q}_0$, whence $\nu(a) \le \mu(a)$ by assumption, a clear contradiction as $\nu(a) \ge \mu(a)+\delta$.

Now we are ready to formulate a Korovkin type approximation theorem for locally convex cones:

1.11 Convergence Theorem. *Let Q_0 be a subcone of the locally convex cone (Q,V). Let $K \subset Q^*$ be a strictly separating set of linear functionals. For any net of u-equicontinuous linear operators $T_\alpha : Q \to Q$ one has:*

(i) $T_\alpha(b) \uparrow b$ *for all* $b \in Q_0$ *implies* $T_\alpha(a) \uparrow a$ *for all elements* $a \in Q$ *that are* Q_0*-superharmonic on* K.

(ii) $T_\alpha(b) \downarrow b$ *for all* $b \in Q_0$ *implies* $T_\alpha(a) \downarrow a$ *for all elements* $a \in Q$ *that are* Q_0*-subharmonic on* K.

(iii) $T_\alpha(b) \to b$ *for all* $b \in Q_0$ *implies* $T_\alpha(a) \to a$ *for all elements* $a \in Q$ *that are* Q_0*-super-and subharmonic on* K.

Similar statements hold for sequences of linear operators and sequential super- and subharmonicity .

We postpone the proof of this convergence theorem. Because of the applications that we have in mind we shall formulate and prove a more general result at the end of this section. Before, let us formulate a slight generalization of the above theorem. As in classical Korovkin theory, 1.9 can be seen as an assertion about pointwise convergence of a net (T_α) of operators towards the identity operator. If we replace the identity by an arbitrary operator $S : Q \to P$ we model 'universal' Korovkin type approximation:

1.12 Convergence Theorem. *Let (Q,V) and (P,W) be locally convex cones and Q_0 a subcone of Q. Let $S : Q \to P$ be a u-continuous linear operator and $K \subset Q^*$ such that $(S^*)^{-1}(K) \subset P^*$ is strictly separating for P. For any net of u-equicontinuous linear operators $T_\alpha : Q \to P$ one has:*

(i) $T_\alpha(b) \uparrow S(b)$ *for all* $b \in Q_0$ *implies* $T_\alpha(a) \uparrow S(a)$ *for all elements* $a \in Q$ *that are* Q_0*-superharmonic on* K.

(ii) $T_\alpha(b) \downarrow S(b)$ *for all* $b \in Q_0$ *implies* $T_\alpha(a) \downarrow S(a)$ *for all elements* $a \in Q$ *that are* Q_0*-subharmonic on* K.

(iii) $T_\alpha(b) \to S(b)$ *for all* $b \in Q_0$ *implies* $T_\alpha(a) \to S(a)$ *for all elements* $a \in Q$ *that are* Q_0*-super- and subharmonic on* K.

Similar statements hold for sequences of linear operators and sequential super- and subharmonicity .

The preceding Convergence Theorems 1.11 and 1.12 suffice to derive the classical results about Korovkin approximation. We shall, however, give a more general version that we shall use in its full strength only in Ch. VI investigating quantitative estimates. The following General Convergence Theorem clearly contains Theorems 1.11 and 1.12 as special cases. One just has to remember that the eventual preorder on nets (see 1.7) includes convergence if one of the nets is stationary.

1.13 General Convergence Theorem. *Let Q_0 be a subcone of the locally convex cone (Q,V). Let $K \subset Q^*$ and $a \in Q$ such that $\mu(a) < +\infty$ for all $\mu \in K$.*

(i) *The element a is Q_0-superharmonic on K if and only if the following holds: For every locally convex cone (P,W) and every pair $(T_\alpha, S_\alpha)_{\alpha \in A}$ of u-equicontinuous linear operators from Q into P one has:*

$$(T_\alpha(b)) \precsim (S_\alpha(b)) \text{ for all } b \in Q_0 \text{ implies } (T_\alpha(a)) \precsim_w (S_\alpha(a))$$

for all $w \in W$ such that the subset $\bigcap_{\alpha \in A} (S_\alpha^*)^{-1}(K)$ in P^* is strictly separating relative to w.

(ii) *The element a is Q_0-subharmonic on K if and only if the following holds: For every locally convex cone (P,W) and every pair $(T_\alpha, S_\alpha)_{\alpha \in A}$ of u-equicontinuous linear operators from Q into P one has:*

$$(S_\alpha(b)) \precsim (T_\alpha(b)) \text{ for all } b \in Q_0 \text{ implies } (S_\alpha(a)) \precsim_w (T_\alpha(a))$$

for all $w \in W$ such that the subset $\bigcap_{\alpha \in A} (S_\alpha^*)^{-1}(K)$ in P^* is strictly separating relative to w.

The local preorder $\precsim_w$ in the above statements may be replaced by the global preorder $\precsim$ if the set $\bigcap_{\alpha \in A} (S_\alpha^*)^{-1}(K)$ is strictly separating for Q.

Similar statements hold for sequential super- and subharmonicity and u-equicontinuous sequences of linear operators.

Proof. We shall give the proof for part (i) and for arbitrary nets: Firstly, assume that a is not Q_0-superharmonic on K. Then there are u-equicontinuous nets $(\nu_\alpha)_{\alpha \in A}$ in Q^* and $(\mu_\alpha)_{\alpha \in A}$ in K such that $(\nu_\alpha(b)) \precsim (\mu_\alpha(b))$ for all $b \in Q_0$, but $\nu_\alpha(a)) \geq \mu_\alpha(a)+\rho$ for some $\rho > 0$ and all $\alpha \in A$. If we set $P = \overline{R}$ and $T_\alpha = \nu_\alpha$, $S_\alpha = \mu_\alpha$ then for all $\alpha \in A$ and $1 \in \overline{R}^*$ we have $\mu_\alpha^*(1)(c) = \mu_\alpha(c)$ for all $c \in Q$, whence $\mu_\alpha^*(1) = \mu_\alpha \in K$ and $1 \in \bigcap_{\alpha \in A} \mu_\alpha^{*-1}(K)$. But the subset $\{1\} \subset \overline{R}^*$ clearly supports the strict separation property for $\overline{R}$. So the property in our theorem does not hold in this case.

Now assume that a is Q_0-superharmonic on K and let (P,W), $(T_\alpha, S_\alpha)_{\alpha \in A}$, and $w \in W$ be as in (i). Furthermore, assume that $(S_\alpha(a)) \precsim_w (T_\alpha(a))$ does not hold in this case; i.e. there are $\delta > 0$ and subnets $(T_\beta, S_\beta)_{\beta \in B}$ such that

$$T_\beta(a) \not\precsim S_\beta(a)+\delta w.$$

By our assumptions on w and on $\bigcap_{\alpha \in A} S_\alpha^{*-1}(K)$ there is a u-equicontinuous net of functionals $\tau_\beta \in P^*$, (i.e. $\tau_\beta \in u^\circ$ for some $u \in W$), and $\rho > 0$ such that

$$\tau_\beta(T_\beta(a)) \geq \tau_\beta(S_\beta(a))+\rho \text{ and } \mu_\beta = S_\beta^*(\tau_\beta) \in K.$$

Now we set $\nu_\beta = T_\beta^*(\tau_\beta)$ and observe that both nets $(\nu_\beta)_{\beta \in B}$ and $(\mu_\beta)_{\beta \in B}$ are u-equicontinuous in Q^* (Remark 1.2(b)). Then, given $b \in Q_0$ there is some $\beta_0 \in B$ such that $T_\beta(b) \leq S_\beta(b)+u$ for all $\beta \geq \beta_0$, whence

$$\nu_\beta(b) = T_\beta^*(\tau_\beta)(b) = \tau_\beta(T_\beta(b)) \leq \tau_\beta(S_\beta(b))+1 = \mu_\beta(b)+1.$$

But this shows that $(\nu_\beta(b)) \leq (\mu_\beta(b))$ for all $b \in Q_0$ (see Remark 1.5(a)), whence $(\nu_\beta(a)) \leq (\mu_\beta(a))$ by our assumption on a. This yields an obvious contradiction to

$$\nu_\beta(a) = T_\beta^*(\tau_\beta)(a) = \tau_\beta(T_\beta(a)) \geq \tau_\beta(S_\beta(b)) + \rho = \mu_\beta(b) + \rho$$

and completes our proof.

2. Some classical applications.

Combining the Convergence Theorems 1.11 and 1.12 (which are in fact corollaries of our General Convergence Theorem 1.13) with the results of Ch. III, the classical versions of Korovkin's Theorem for positive linear operators on $C(X)$ or, more generally, on locally convex topological vector lattices are now immediately available. Those cases have been dealt with in many places (c.f. [12], [19], [20], [49], etc.), and we do not intend to repeat them in detail. We shall, however, give a few examples that may indicate how to apply our general results to various classical situations. First, let us recall some facts and comment on the three hypotheses in 1.11 and 1.12: (a) u-equicontinuity, (b) strict separation, and (c) superharmonicity.

2.1 Remarks. (a) Suppose that (Q,V) is a full cone (see I.2.5) and that V is contained in the subcone Q_0. If we have a net of linear operators $T_\alpha : Q \to Q$ such that $T_\alpha(b) \uparrow b$ for all $b \in Q_0$, then the net $(T_\alpha)_{\alpha \in A}$ is eventually u-equicontinuous: Indeed, for every $v \in V$ we have $T_\alpha(v) \uparrow v$, i.e. there is an $\alpha_0 \in A$ such that $T_\alpha(v) \leq 2v$ for all $\alpha \geq \alpha_0$; thus $a \leq b + v$ implies $T_\alpha(a) \leq T_\alpha(b) + 2v$ for all $\alpha \geq \alpha_0$. Thus, for Theorems 1.11, 1.12 and 1.13 we can replace the hypothesis of u-equicontinuity of the T_α by the simpler property of being order preserving, if Q_0 contains all of V.

(b) The subset $K \subset Q^*$ on which we have to check the superharmonicity of elements $a \in Q$ should be as small as possible. On the other hand, K has to be big enough in order to be strictly separating for Q. For the latter it is sufficient that K contains (suitable multiples of) the extreme points of all polars v_Q°, $v \in V$, provided that Q is tightly covered by its bounded elements (see Proposition 1.5).

(c) The property for an element $a \in Q$ to be Q_0-superharmonic on K according to Definition 1.8 is unpleasant. If K intersects the polar of every neighborhood in a $w(Q^*,Q)$-closed set, in particular, if K itself is $w(Q^*,Q)$-closed, then by Proposition 1.10 we may replace this hypotheses by: a is Q_0-superharmonic in every $\mu \in K$ according to Definition III.1.1, i.e for all $\nu \in Q^*$ and $\mu \in K$ we have $\mu(a) < +\infty$, and

$$\nu(b) \leq \mu(b) \text{ for all } b \in Q_0 \text{ implies } \nu(a) \leq \mu(a).$$

Now we may also use the information about superharmonic elements collected in Chapter III. For example, by the Sup-Inf-Theorem III.1.3, a is Q_0-superharmonic in μ if and only if

$$\mu(a) = \sup_{v \in V} \inf\{\mu(b) \mid b \in Q_0,\ a \leq b+v\}.$$

In case that $V \subset Q_0$, by II.1.4 this last equation may be further simplified

$$\mu(a) = \inf\{\mu(b) \mid b \in Q_0,\ a \leq b\}.$$

In the classical situations of Korovkin approximation we may always choose K to be a $w(Q^*,Q)$-closed subset of Q^*. In Chapter VI, however, we will study situations where this will no longer be feasible, and we will have to handle the general notion of superharmonicity as in 1.8.

The first of the following two exemplary propositions is due to Bauer and Donner [11]. (In fact, they prove a more general result for locally compact spaces X and the space $C_0(X)$ of continuous functions on X vanishing at infinity. This situation will be covered and generalized by our Proposition 2.1 in Ch. V.) The second one does not seem to be in the literature in this explicit form, although it can be derived from more general results as in [3] or [4].

2.2 Proposition. *Let X be a compact Hausdorff space and G a linear subspace of $C(X)$. For a function $f \in C(X)$ the following conditions are equivalent:*

(i) For every net $(T_\alpha)_{\alpha \in A}$ of equicontinuous positive linear operators on $C(X)$

$$T_\alpha(g) \to g \text{ for all } g \in G, \text{ implies } T_\alpha(f) \to f.$$

(ii) For every $x \in X$ we have

$$f(x) = \sup_{\varepsilon>0} \inf\{g(x) \mid g \in G,\ f \leq g+\varepsilon\} = \inf_{\varepsilon>0} \sup\{g(x) \mid g \in G,\ g \leq f+\varepsilon\}.$$

(iii) For every $x \in X$ and every positive Radon measure μ on X

$$\mu(g) = g(x) \text{ for all } g \in G, \text{ implies } \mu(f) = f(x).$$

Convergence in (i) is meant with respect to the topology of uniform convergence on $C(X)$.

Proof. We consider the locally convex cone $(C(X),V)$ with its usual neighborhoods $V = \{\rho > 0\}$. Then clearly, (iii) means that f is G-sub- and superharmonic in all point evaluations $\varepsilon_x \in C(X)^*$ which by the Sup-Inf-Theorem III.1.3 is equivalent to (ii). Furthermore, if we set $K = \{\varepsilon_x \mid x \in X\}$ the Convergence Theorem 1.11 shows that (iii) implies (i). Note that for linear operators on $C(X)$ u-equicontinuity means positivity and equicontinuity in the usual sense. So all left to show is that (i) implies (iii) as well. The following argument uses specific properties of $C(X)$ and does not hold in the general case which by the General Convergence Theorem 1.13 requires the consideration of arbitrary nets (S_α) in order to yield this equivalence: Suppose by the way of contradiction that for $f \in C(X)$ there are $x \in X$ and a positive Radon measure μ on X such that $\mu(g) = g(x)$ for all $g \in G$, but $\mu(f) \neq f(x)$. For every open neighborhood U of x choose $\phi_U \in C(X)$ such that

$$0 \leq \phi_U \leq 1,\ \phi_U(x) = 1, \text{ and } \phi_U|_{X \setminus U} = 0.$$

Now define the operator T_U by

$$T_U(h) = h(1-\phi_U) + \mu(h)\,\phi_U \text{ for all } h \in C(X).$$

Clearly, T_U is a positive linear operator on $C(X)$. As $\|T_U\| \leq 1+\|\mu\|$, the T_U form an equicontinuous net as U runs through a neighborhood basis of x. As $T_U(h)-h = (\mu(h)-h)\phi_U$, one easily verifies that $T_U(g) \to g$ for all $g \in G$, but $T_U(f)$ is not convergent; i.e. (i) does not hold for f, thus completing our proof.

A corresponding Korovkin type theorem for contractive linear operators on normed spaces may be derived using our Example II.2.17 and the subsequent remarks on it. By $C_{\mathbb{C}}(X)$ we denote the space of continuous complex-valued functions on the compact space X.

2.3 Proposition. *Let X be a compact Hausdorff space and G a (complex linear) subspace of $C_{\mathbb{C}}(X)$. For a function $f \in C_{\mathbb{C}}(X)$ the following conditions are equivalent:*

(i) *For every net $(T_\alpha)_{\alpha \in A}$ of contractive linear operators on $C_{\mathbb{C}}(X)$.*

$$T_\alpha(g) \to g \quad \text{for all } g \in G, \quad \text{implies } T_\alpha(f) \to f.$$

(ii) *For every $x \in X$ and for all $\gamma \in \mathbb{C}$ such that $|\gamma| = 1$ we have*

$$\mathrm{Re}(\gamma f(x)) = \inf\{\mathrm{Re}(\gamma g(x)) + \|f-g\| \mid g \in G\}.$$

(iii) *For every $x \in X$ and every complex Radon measure μ on X such that $\|\mu\| \le 1$*

$$\mu(g) = g(x) \quad \text{for all } g \in G, \quad \text{implies } \mu(f) = f(x).$$

Convergence in (i) is meant with respect to the topology of uniform convergence on $C_{\mathbb{C}}(X)$.

Proof. Let (Q,V) be the full locally convex cone as in our Example II.2.17; i.e. $Q = \{f+\rho B \mid f \in C_{\mathbb{C}}(X),\ \rho \ge 0\}$ and $V = \{\rho B \mid \rho > 0\}$, where B denotes the unit ball of $C_{\mathbb{C}}(X)$. Q is ordered by inclusion. The dual cone Q^* may be identified with the set of all elements $\mu \oplus r$, where μ denotes a complex Radon measure on X of norm at most r, and the functional operates on Q as follows:

$$(\mu \oplus r)(f+\rho B) = \mathrm{Re}(\mu(f)) + r\rho.$$

The set $K = \{\gamma\varepsilon_x \oplus 1 \mid x \in X,\ |\gamma| = 1\}$ is seen to be $w(Q^*,Q)$-compact and supports the strict separation property for Q. Every continuous linear operator T on $C_{\mathbb{C}}(X)$ may be canonically extended to a u-continuous linear operator on Q by setting $T(B) = \|T\|B$. We investigate superharmonicity with respect to the subcone $Q_0 = \{g+\rho B \mid g \in G,\ \rho \ge 0\}$. Note that convergence towards the identity on Q_0 requires our operators to be (almost) contractive (in fact, we only need that $\lim\sup_\alpha \|T_\alpha\| \le 1$). Condition (iii) of the proposition means that the function f (and all its complex multiples) is Q_0-superharmonic in the elements of K. Moreover, Q_0 contains the neighborhoods of Q, and the equivalence of (ii) and (iii) is an immediate consequence of Corollary III.1.4 to the Sup-Inf-Theorem III.1.3: The function f is Q_0-superharmonic in $\gamma\varepsilon_x \oplus 1$ if and only if

$$\begin{aligned}\mathrm{Re}(\gamma\varepsilon_x(f)) &= \inf\{\mathrm{Re}(\gamma\varepsilon_x(g)) + \rho \mid g \in G,\ f \le g+\rho B\}\\ &= \inf\{\mathrm{Re}(\gamma g(x)) + \|f-g\| \mid g \in G\}.\end{aligned}$$

The Convergence Theorem 1.11 shows that (iii) implies (i), and the converse implication follows by the same construction of a net of operators T_U as in the proof of the preceding corollary. Those operators are seen to be linear and contractive in this case.

2.4 Remarks. (a) If the subspace G contains the constant functions, then the respective conditions (iii) in Propositions 2.2 and 2.3 (whence conditions (i) and (ii) as well) coincide: For the complex Radon measure μ on X, the conditions $\mu(1) = 1$ and $\|\mu\| \leq 1$ imply already that μ is positive.

(b) In 'universal' Korovkin type approximation theory the operators under consideration have ranges different from their domains. Using our General Convergence Theorem 1.13 , in Propositions 2.2 and 2.3 the identity operator might be replaced by a more general net of operators $(S_\alpha)_{\alpha \in A}$ with ranges in spaces $C(Y)$ resp. $C_{\mathbb{C}}(Y)$ for some compact space Y. Suitable operators S_α might be lattice homomorphisms in the first case and surjective isometries in the second one (c.f. Ch. II, Remark 6.2(c) and Example 6.10(d)), because they are seen to satisfy the condition on the S_α from Theorem 1.13. Furthermore, it is well-known that the adjoint operators of algebra homomorphisms from $C_{\mathbb{C}}(X)$ into $C_{\mathbb{C}}(Y)$ map point evaluations on Y into point evaluations on X as well. Thus, in the case of Proposition 2.3 we may also let the operators S_α be non-zero algebra homomorphisms. We shall use this last observation in our Example 2.5.

(c) The equivalence of conditions (ii) and (iii) in the above propositions still holds if we consider subspaces, i.e. function spaces instead of the whole spaces $C(X)$ resp. $C_{\mathbb{C}}(X)$. In order to study convergence for suitable nets of operators in this case, i.e. for the implication from (iii) to (i), it suffices to consider superharmonicity in the point evaluations of the Silov boundary of those function spaces, because this set is strictly separating. Korovkin type theorems have for instance been studied in [21]. In the universal case and for positive operators the suitable operators S_α are just our directional operators as introduced in II.6. They are called R_ε-homomorphisms in [21] and defined as follows (this is a special version of our Definition II.6.1): A positive linear operator S between two function spaces E and F is called an *R_ε-homomorphism* if, for all $f,g \in E$ and $h \in F$ such that both $S(f) \leq h$ and $S(g) \leq h$, and for each $\varepsilon > 0$ we can find a function $d \in E$ such that $d \leq f$, $d \leq g$ and $S(d) \leq h+\varepsilon$. For subalgebras of $C_{\mathbb{C}}(X)$, i.e. function algebras, the suitable operators S_α are algebra homomorphisms (c.f. 1.13 (b)).

In order to illustrate the versatility of our approach we shall, in the following, give some further examples dealing with less known situations. Typically, we consider certain restricted classes of continuous linear operators on locally convex vector spaces. We take advantage of the fact that for locally convex cones those restrictions may be taken care of by the proper choice of our domains and ranges and their topologies alone.

2.5 Example: Approximation in commutative Banach algebras. The following is due to F. Altomare [4] and may be derived using our Proposition 2.3: Let E be a commutative complex Banach algebra with identity element e. We denote by ΔE the compact set of all multiplicative linear functionals in the dual E' of E, and for $f \in E$ by $\hat{f}$ its Gelfand transform as a continuous complex-valued function on ΔE. As usual, $\|f\|^\wedge$ denotes the spectral

norm of f, i.e. the supremum norm of $\hat{f}$ on ΔE. Altomare investigates Korovkin type theorems for contractive linear operators with respect to this (semi-) norm. By $\hat{E}$ let us denote the image of E under the Gelfand transform as a subspace of $C_{\mathbb{C}}(\Delta E)$. It is clear then that the operators considered by Altomare are just contractions on this subspace $\hat{E}$ of $C_{\mathbb{C}}(\Delta E)$, and we may apply Proposition 2.3 together with Remark 2.4 (c):
Let G be a subspace of E which contains the identity e and $\hat{G}$ the corresponding function space of its Gelfand transforms. The function $\hat{f} \in \hat{E}$ then is $\hat{G}$-superharmonic in all $\mu \in \Delta E$ if and only if for all $\mu \in \Delta E$ and all probability measures ν on ΔE such that $\nu(\hat{g}) = \mu(g)$ for all $g \in G$, we have $\nu(\hat{f}) = \mu(f)$ as well.
Now let F be another commutative Banach algebra and $(T_\alpha, S_\alpha)_{\alpha \in A}$ a double net of linear operators from E into F. We choose all operators to be contractive (with respect to the spectral norms on E and F) and the S_α to be non-zero algebra homomorphisms in addition. Then Proposition 2.3 immediately yields Altomare's result:

For $f \in E$ such that the corresponding function $\hat{f} \in \hat{E}$ then is $\hat{G}$-superharmonic in all $\mu \in \Delta E$ and all such nets $(T_\alpha, S_\alpha)_{\alpha \in A}$ of operators as above we have

$$\|T_\alpha(f) - S_\alpha(f)\|^\wedge \to 0, \quad \text{whenever} \quad \|T_\alpha(g) - S_\alpha(g)\|^\wedge \to 0 \ \text{ for all } g \in G.$$

2.5 Example: Approximation in locally convex vector spaces. A very general setting for Korovkin type classes of linear operators in this case is introduced in [49]: Let E be a locally convex topological vector space with 0-neighborhood base V. Let U be a cone of convex subsets of E and consider equicontinuous nets of linear operators $(T_\alpha)_{\alpha \in A}$ on E such that for every $U \in \boldsymbol{U}$ and $V \in \boldsymbol{V}$ there is $\alpha_0 \in A$ such that

$$T_\alpha(U) \subset U + V \quad \text{for all } \alpha \geq \alpha_0.$$

The aim is to describe those elements $e \in E$ such that $\lim_\alpha T_\alpha(e) = e$. This situation can be easily modelled using suitable locally convex cones: Let $(\boldsymbol{E}, \boldsymbol{V})$ be a locally convex cone and $(\boldsymbol{Q}, \overline{\boldsymbol{V}})$ a subcone of $(\overline{DConv}(\boldsymbol{E}), \overline{\boldsymbol{V}})$ as introduced in Ch. I, 2.8. Assume that $\boldsymbol{Q}$ contains all singleton subsets of $\boldsymbol{E}$ and let $\boldsymbol{Q}_0$ be a subcone of $\boldsymbol{Q}$. (To model the situation in [49] let $\boldsymbol{E} = E$, $\boldsymbol{V}$ be as above, $\boldsymbol{Q} = \{\{e\} + U \mid e \in \boldsymbol{E},\ U \in \boldsymbol{U}\}$, and $\boldsymbol{Q}_0 = \boldsymbol{U}$.) As the dual cone $\boldsymbol{Q}^*$ tends to be hard to handle we restrict ourselves to the canonical embedding

$$\mu \to \overline{\mu} : \boldsymbol{E}^* \to \boldsymbol{Q}^* \quad \text{defined by} \quad \overline{\mu}(A) = \sup\{\mu(a) \mid a \in A\} \ \text{ for } A \in \boldsymbol{Q}.$$

If all elements of $\boldsymbol{E}$ are bounded then $\boldsymbol{E}^*$ supports the separation property for $\boldsymbol{Q}$ (c.f. II.2.16). Now let K be a subset of $\boldsymbol{E}^*$ such that its intersection with the polar of each neighborhood of $\boldsymbol{E}$ is $w(\boldsymbol{E}^*, \boldsymbol{E})$- closed, whence $w(\boldsymbol{E}^*, \boldsymbol{E})$-compact. If for all $A \in \boldsymbol{Q}$ and all $v \in \boldsymbol{V}$ the mapping $\mu \to \overline{\mu}(A) : K \cap v_{\boldsymbol{E}}^{\circ} \to \overline{\boldsymbol{R}}$ is continuous (with respect to $w(\boldsymbol{E}^*, \boldsymbol{E})$) then the embedding $\overline{K}$ of K into $\boldsymbol{Q}^*$ is immediately seen to have the same property as K: Its intersection with the polar of each neighborhood of $\boldsymbol{Q}$ is $w(\boldsymbol{Q}^*, \boldsymbol{Q})$-compact. This property was seen to be useful because it considerably facilitates the investigation of super- and subharmonicity (Remark 2.1(c)). Now we may apply the General Convergence Theorem 1.13 with any locally convex cone $(\boldsymbol{P}, \boldsymbol{W})$ and suitable nets (T_α, S_α) of linear operators and

neighborhoods $w \in W$. If in particular we set $P = \overline{DConv(E)}$, and all the S_α equal the identity operator, then we are in the situation of [49].
Note that the continuity of the mapping

$$\mu \to \overline{\mu}(A) \ : \ K \cap v_E^\circ \ \to \overline{\mathbb{R}}$$

is in particular guaranteed if the element $A \in Q$, as a subset of E, is precompact with respect to the upper topology on E: Indeed, for $\varepsilon > 0$ there are elements $a_1, a_2, \ldots a_n \in A$ such that the upper neighborhoods $(\varepsilon v)(a_1), (\varepsilon v)(a_2), \ldots, (\varepsilon v)(a_n)$ cover all of A. But this shows that the continuous function on $K \cap v_E^\circ$, $\mu \to (\mu(a_1) \vee \ldots \vee \mu(a_n))$ approximates $\mu \to \overline{\mu}(A)$ up to ε, whence the the latter mapping is continuous as well.

Let us illustrate this situation with a concrete example: Let $E = C[0,1]$ with its usual sup-norm topology (i.e. as a locally convex cone, E carries the trivial order). Let B denote the unit ball of E and F and G the subspaces

$$F = \left\{ f \in E \mid f(0) = 0 \right\}, \quad G = \left\{ \text{polygonal arcs } g \text{ satisfying } g(0) = \int_0^1 g(x)\, dx \right\}.$$

Let $Q = \{\{e\} + \gamma F + \rho B \mid e \in E, \gamma, \rho \geq 0\}$ and $Q_0 = \{\{g\} + \gamma F + \rho B \mid g \in G, \gamma, \rho \geq 0\}$. For $0 < x \leq 1$ now consider the point evaluation ε_x and its embedding $\overline{\varepsilon}_x$ into Q^*. For every $\overline{\mu} \in Q^*$ such that $\overline{\mu}(b) \leq \overline{\varepsilon}_x(b)$ for all $b \in Q_0$ w e conclude, firstly, that $\overline{\mu}(B) \leq \overline{\varepsilon}_x(B) = 1$. Thus, $\overline{\mu}$ coincides on E with a Radon measure μ on $[0,1]$ of norm at most 1. Secondly, as all constant functions are contained in Q_0 we have $\overline{\mu}(1) = \mu(1) = 1$. This yields that μ is in fact a probability measure on $[0,1]$. But since $G \subset Q_0$ clearly contains a positive function which vanishes only in x we conclude, thirdly, using the support of this measure, that $\mu = \varepsilon_x$. For $\overline{\varepsilon}_0$ and $\overline{\mu} \in Q^*$ such that $\overline{\mu}(b) \leq \overline{\varepsilon}_0(b)$ for all $b \in Q_0$ we conclude in a similar way from $F \subset Q_0$ that the underlying Radon measure μ coincides on E with ε_0. Thus indeed, every function $e \in E$ is both Q_0- super- and subharmonic in all point evaluations. But the set $\overline{K} = \{\overline{\varepsilon}_x \mid 0 \leq x \leq 1\}$ is no longer $w(Q^*, Q)$-compact as the mapping $\varepsilon_x \to \overline{\varepsilon}_x(F)$ is discontinuous in ε_0. However, for all $0 < \rho \leq 1$ we clearly have continuity on the interval $[\rho,1]$, whence the sets $\overline{K}_\rho = \{\overline{\varepsilon}_x \mid \rho \leq x \leq 1\}$ are all seen to be $w(Q^*, Q)$-compact. Finally, for the neighborhoods $w_\rho = \{e \in E \mid e(x) \leq 1 \text{ for all } \rho \leq x \leq 1\}$ the strict separation property is clearly supported by the sets K_ρ. Now we may apply the Convergence Theorem 1.11: A contractive linear operator $T : E \to E$ is extended to a u-continuous linear operator $\overline{T} : Q \to Q$ by setting $\overline{T}(B) = B$ and $\overline{T}(f) = F$. Using both the super- and the subharmonicity part of our theorem with the neighborhoods w_α we conclude that for all nets (T_α) of such operators which converge to the identity for all $g \in G$ then $T_\alpha(e)$ converges to e uniformly on the intervals $[\rho,1]$ and in 0 for all $e \in E$.

The following specific counterexample, however, will show that uniform convergence on the whole interval $[0,1]$ is in general not attainable for this example:

For $e \in C[0,1]$ set $\rho = \int_0^1 e(x)\, dx$, and define $T_n(e)$ by

$$T_n(e)(x) = \begin{cases} \text{line connecting } \left(0, e(0)\right) \text{ and } \left(\frac{1}{n}, \rho\right), & \text{for } 0 \le x \le \frac{1}{n}, \\ \text{line connecting } \left(\frac{1}{n}, \rho\right) \text{ and } \left(\frac{2}{n}, e(\frac{2}{n})\right), & \text{for } \frac{1}{n} \le x \le \frac{2}{n}, \\ e(x), \text{ otherwise.} \end{cases}$$

It is obvious that these operators are linear and contractive and converge towards the identity operator on G. But for $e \in C[0,1]$, $T_n(e)$ converges to e uniformly on the whole interval $[0,1]$ if and only if

$$\int_0^1 e(x)\, dx = e(0).$$

A further inspection of this example reveals that the closure of $\overline{K}$ in Q^* (which as an equicontinuous subset of Q^* is $w(Q^*,Q)$-compact) contains in addition the functional $\tilde{\varepsilon}_0$ such that

$$\tilde{\varepsilon}_0(e) = e(0) \quad \text{for all} \quad e \in C[0,1], \quad \tilde{\varepsilon}_0(B) = 1, \quad \text{and } \tilde{\varepsilon}_0(F) = +\infty.$$

Subharmonicity and superharmonicity for $e \in C[0,1]$ in $\tilde{\varepsilon}_0$, however, yield exactly the above condition $\int_0^1 e(x)\, dx = e(0)$.

•

Chapter V: Nachbin Cones

In this chapter we are going to apply our general results of Chapter IV, in particular the Convergence Theorems IV.1.11 and 1.13, in order to obtain new Korovkin type theorems. We consider spaces of continuous functions defined on locally compact Hausdorff spaces X with values in locally convex cones Q. The main idea is to build Korovkin systems of cone-valued functions from Korovkin systems of real-valued functions. As we want to include weighted approximation in our considerations, we first introduce Nachbin cones in Section 1. These are the obvious generalizations of Nachbin spaces, i.e. vector spaces of continuous functions defined on X with values in a locally convex vector space, where the topology on the function space is ruled by a family of weight functions. For details on Nachbin spaces we globally refer to Prolla [43]. Some early Korovkin type results on the weighted approximation of real-valued functions in one and several variables with a single weight function are due to Gadzhiev [23], [24]. In order to apply our general results to Nachbin cones, we have to provide some information about their duals.

In Section 2 we first present an analogue of Proposition IV.2.2 for weighted spaces of real-valued continuous functions. Then we prove the main Theorem 2.5 of this section which is not only important but, unfortunately, very technical, too. As a remedy we offer a simpler (and less powerful) version which, however, proves sufficient for some cases (Theorem 2.4). The application to the approximation of stochastic processes in Corollary 2.8 shows that the effort was worthwhile.

In Section 3, finally, we apply Theorems 2.4 and 2.5 to weighted spaces of set-valued functions. Our results generalize and cover previous results by Vitale [59] and the authors [28]. They are closesly related to recent work by Campiti [13], [14].

Throughout this chapter let

X be a locally compact Hausdorff space,

(Q,V) a locally convex cone,

$C_s(X,Q)$ the cone of all functions $f: X \to Q$ which are continuous with respect to the symmetric topology on Q (c.f. Chapter I, Example 2.9).

1. Weighted cones of continuous cone-valued functions.

In Chapter I, Example 2.9 we introduced the cone $C_s(X,Q)$ of Q-valued functions on X which are continuous with respect to the symmetric topology of Q. On $C_s(X,Q)$ we considered the topology of uniform convergence on X induced by the neighborhoods in Q. We shall now generalize this concept by introducing weight-functions, thus defining Nachbin cones of cone valued functions.

1.1 Weight functions. A family Ψ of upper semicontinuous non-negative real-valued functions on the locally compact space X is said to be *directed upward* if for $\psi, \varphi \in \Psi$ there is some $\phi \in \Psi$ and $\rho > 0$ such that $\psi(x) \leq \rho\phi(x)$ and $\varphi(x) \leq \rho\phi(x)$ holds for all $x \in X$. Any such family Ψ is called a family of *weight functions* on X. In the following, let Ψ be a fixed family of weights.

1.2 Nachbin cones. Let (Q,V) be a locally convex cone and Ψ a family of weights on the locally compact space X. For every $v \in V$ and $\psi \in \Psi$ let v_ψ be the neighborhood for $C_s(X,Q)$ defined for functions $f, g \in C_s(X,Q)$ by

$$f \leq g + v_\psi \quad \text{if and only if} \quad \psi(x)f(x) \leq \psi(x)g(x) + v \quad \text{for all } x \in X.$$

Then those neighborhoods induce a convex quasiuniform structure on $C_s(X,Q)$ in the sense of Ch. I,5, and the set $\{v_\psi \mid v \in V,\ \psi \in \Psi\}$ generates (i.e. is a base of) an abstract neighborhood system on $C_s(X,Q)$ which we shall denote by V_Ψ.
The *Nachbin cone* $C\Psi(X,Q)$ is the subcone of $C_s(X,Q)$ consisting of those functions $f \in C_s(X,Q)$ which vanish at infinity, i.e. such that for every $v \in V$ and $\psi \in \Psi$ there is a compact subset Y of X such that

$$\psi(x)f(x) \leq v \quad \text{and} \quad 0 \leq \psi(x)f(x) + v \quad \text{for all } x \in X \setminus Y.$$

Note that all such functions are bounded below with respect to the neighborhoods in V_Ψ: Indeed, let $f \in C\Psi(X,Q)$, $v \in V$ and $\psi \in \Psi$. Choose the compact subset Y of X such that the above condition holds. A simple compactness argument shows that the function f, as it is pointwise bounded below, is even uniformly bounded below on Y, i.e. there is $\rho > 0$ such that

$$0 \leq f(x) + \rho v \quad \text{for all } x \in Y.$$

Multiplying this inequality with $\psi(x)$ and considering that the upper semicontinuous function ψ is bounded by some $\lambda > 0$ on Y, this yields

$$0 \leq \psi(x)f(x) + \rho\psi(x)v \leq \psi(x)f(x) + \lambda\rho v \quad \text{for all } x \in Y.$$

Thus, indeed, $0 \leq \psi(x)f(x) + \max\{\lambda\rho, 1\}v$ for all $x \in X$, and $(C\Psi(X,Q), V_\Psi)$ becomes a locally convex cone.
By $C\Psi(X)$ we shall in the following abbreviate the Nachbin cone $C\Psi(X,\mathbf{R})$, where $\mathbf{R}$ is endowed with its usual cone topology generated by the neighborhoods $V = \{\rho > 0\}$.

1.3 Remarks. (a) If Y is a compact subset of X and if all the values of the function $f \in C_s(X,Q)$ are bounded on Y, then an argument similar to the above shows that f and all functions ψf, where ψ denotes any weight function on X, are even uniformly bounded on Y. Furthermore, this observation yields that every function f in the Nachbin cone $C_s(X,Q)$ which assumes only bounded values is bounded as an element of the locally convex cone $(C\Psi(X,Q), V_\Psi)$. On the other hand, if the function f is bounded as an element of $(C\Psi(X,Q), V_\Psi)$, then its values are seen to be bounded in all $x \in X$ such that $\psi(x) \neq 0$ for at least one weight function $\psi \in \Psi$.

(b) For an element $a \in Q$ consider the subcone of Q

$$B_a = \{b \in Q \mid \text{for all } v \in V \text{ there are } \lambda, \rho \geq 0 \text{ such that } b \leq \lambda a + \rho v\}.$$

It is easily checked that B_a is open in the upper and closed in the lower topology of Q, whence both open and closed in the symmetric topology. Thus, if Y is any connected subset of X and $y \in Y$, then we have $f(z) \in B_{f(y)}$ for all $f \in C_s(X,Q)$ and $z \in Y$. In particular, if $f(y)$ is bounded for some $y \in Y$, then all the values of f are bounded on Y. Furthermore, if the whole space X is connected and at least one of the weight functions $\psi \in \Psi$ has a non-compact support, then the above shows, together with the condition in 1.2, that for every function f in the Nachbin cone $C\Psi(X,Q)$ all of its values are bounded in Q; whence (a) applies, and all functions in $C\Psi(X,Q)$ are seen to be bounded.

1.4 Examples. (a) If the family of weights consists only of the constant function 1, then $C\Psi(X,Q)$ is the cone of continuous Q-valued functions which vanish at infinity. As usual, we shall denote it by $C_0(X,Q)$. It is endowed with the topology of uniform convergence on X.

(b) If $\Psi = C_0(X) = C_0(X,R)$ then $C\Psi(X,Q)$ consists of the cone of continuous Q-valued functions which are bounded at infinity; i.e. for every $v \in V$ there is a compact subset Y of X and some $\rho > 0$ such that

$$f(x) \leq \rho v \quad \text{and} \quad 0 \leq f(x) + \rho v \quad \text{for all } x \in X.$$

(c) If Ψ consists of the characteristic functions of all compact subsets of X, then $C\Psi(X,Q) = C_s(X,Q)$ endowed with the topology of uniform convergence on compact subsets of X.

(d) If Ψ consists of the characteristic functions of all finite subsets of X then, again, $C\Psi(X,Q) = C_s(X,Q)$, now endowed with the topology of pointwise convergence on X.

1.5 Linear functionals on Nachbin cones. Looking for the dual cone of $C\Psi(X,Q)$, the case $Q = R$ is fairly easy to handle: $C\Psi(X)^*$ coincides with $\Psi M_B(X)^+$, where $M_B(X)^+$ denotes the set of all positive bounded Radon measures on X, and

$$\Psi M_B(X)^+ = \{\psi\mu \mid \psi \in \Psi,\ \mu \in M_B(X)^+\}.$$

(As we use only the one-sided topology on $C\Psi(X)$, the dual cone consists of positive functionals only. If we would use the symmetric topology, then $M_B(X)^+$ has to be replaced by the set of all Radon measures. For a proof one may consult [43], Theorem 5.42.)

The general case is hard to handle (compare [43]), even if the cone $\mathcal{Q}$ and its dual are well-understood. But in order to apply the Convergence Theorem IV.1.13 we only need to identify sufficiently large subsets of $C\Psi(X,\mathcal{Q})^*$ with the properties required there. We shall proceed to exhibit such subsets:

For $\mu \in \mathcal{Q}^*$ and the point $x \in X$ we denote by μ_x the linear functional on $C\Psi(X,\mathcal{Q})$ defined by

$$\mu_x(f) = \mu(f(x)) \quad \text{for all } f \in C\Psi(X,\mathcal{Q}).$$

1.6 The set $K_{\Psi X}$. For any subset K of $\mathcal{Q}^*$, let $K_{\Psi X}$ be the set of all functionals of the form μ_x with $\mu \in K$ and $x \in X$ which are u-continuous on the Nachbin cone $C\Psi(X,\mathcal{Q})$. In general, μ_x need not be u-continuous. It is u-continuous iff it belongs to some polar v_ψ°. Let us characterize this property:

1.7 Lemma. *Let $\mu \in \mathcal{Q}^*$, $x \in X$ and $\psi \in \Psi$. Then*

$$\mu_x \in v_\psi^\circ \quad \text{iff} \quad \begin{cases} \mu \in \rho v_{\mathcal{Q}}^\circ \ \text{ for all } \rho > \psi(x), & \text{in case } x \in \operatorname{supp}\psi, \\ \mu = 0, & \text{in case } x \notin \operatorname{supp}\psi. \end{cases}$$

More precisely, we shall show the following statements which are easily seen to be equivalent to the lemma:

(i)	*In case* $\psi(x) \neq 0$:	$\mu_x \in v_\psi^\circ$ *iff* $\mu \in \psi(x) v_{\mathcal{Q}}^\circ$.
(ii)	*In case* $\psi(x) = 0$ *but* $x \in \operatorname{supp}\psi$:	$\mu_x \in v_\psi^\circ$ *iff* $\mu \in \rho v_{\mathcal{Q}}^\circ$ *for all* $\rho > 0$.
(iii)	*In case* $x \notin \operatorname{supp}\psi$:	$\mu_x \in v_\psi^\circ$ *iff* $\mu = 0$.

Note that, if every element of $\mathcal{Q}$ is bounded, then $\mu \in \rho v_{\mathcal{Q}}^\circ$ for all $\rho > 0$ implies $\mu = 0$. Thus, in this case, μ_x is u-continuous if and only if $\psi(x) \neq 0$ for some $\psi \in \Psi$ or $\mu = 0$.

For the proof, let us suppose first that $\mu_x \in v_\psi^\circ$. Let $a,b \in \mathcal{Q}$ such that $a \le b+v$. Given $\rho > \psi(x)$, there is a neighborhood U of x such that $\psi(y) < \rho$ for all $y \in U$. Now, using complete regularity we may find a positive function $f \in C(X)$ with compact support contained in U such that $f(x) = 1/\rho$, and $0 \le f(y) \le 1/\rho$ for all $y \in X$. But then

$$\psi(y)f(y) \le 1, \quad \text{whence} \quad \psi(y)f(y)a \le \psi(y)f(y)b + v \quad \text{for all } y \in X.$$

As the functions $f{\cdot}a$ and $f{\cdot}b$ are contained in $C\Psi(X,\mathcal{Q})$ and as $f{\cdot}a \le f{\cdot}b + v_\psi$, it follows that

$$f(x)\mu(a) = \mu_x(f{\cdot}a) \le \mu_x(f{\cdot}b) + 1 = f(x)\mu(b) + 1,$$

whence $\mu(a) \le \mu(b) + 1/f(x) = \mu(b) + \rho$. The latter shows $\mu \in \rho v_{\mathcal{Q}}^\circ$ for all $\rho > \psi(x)$.

If $\psi(x) > 0$ this implies $\mu \in \psi(x) v_{\mathcal{Q}}^\circ$.

If $x \notin \operatorname{supp}\psi$, then ψ vanishes on a neighborhood U of x. There is a positive function $f \in C(X)$ with compact support contained in U and $f(x) = 1$. For a given $a \in \mathcal{Q}$ the function $f{\cdot}a$ is contained in $C\Psi(X,\mathcal{Q})$ and we have $f{\cdot}a \in \varepsilon v_\psi^\circ$ for all $\varepsilon > 0$. But this shows $\mu(a) = f(x)\mu(a) = 0$, whence $\mu = 0$. This finishes the proof of the necessity of the conditions in (i), (ii), (iii).

For the sufficiency, nothing is to be proved in (iii). For (i) let $\psi(x) \neq 0$ and $\mu \in \psi(x)v_Q^\circ$. Whenever $f \leq g+v_\psi$, then $\psi(x)f(x) \leq \psi(x)g(x)+v$, whence $\psi(x)\mu(f(x)) \leq \psi(x)\mu(g(x))+\psi(x)$, i.e. $\mu_x(f) \leq \mu_x(g)+1$, and we have shown that $\mu_x \in v_\psi^\circ$. For (ii) let $\psi(x) = 0$ but $x \in \operatorname{supp} \psi$, and suppose that $\mu \in \rho v_Q^\circ$ for all $\rho > 0$. Consider any $f,g \in C\Psi(X,Q)$ with $f \leq g+v_\psi$, i.e. $\psi(y)f(y) \leq \psi(y)g(y)+v$ for all $y \in X$. By the symmetric continuity of f and g there is a neighborhood U of x such that both $f(x) \leq f(y)+v$ and $g(y) \leq g(x)+v$ hold for all $y \in U$. Combined, for all $y \in U$ these inequalities yield

$$\psi(y)f(x) \leq \psi(y)f(y)+\psi(y)v \leq \psi(y)g(y)+(\psi(y)+1)v \leq \psi(y)g(x)+(2\psi(y)+1)v\,,$$

whence $\psi(y)\mu(f(x)) \leq \psi(y)\mu(g(x))$. As $\psi(y) \neq 0$ for some $y \in U$, we conclude that $\mu_x(f) \leq \mu_x(g) \leq \mu_x(g)+1$. Thus $\mu_x \in v_\psi^\circ$, indeed.

Next we observe the following:

1.8 Proposition. *Let K be a subset of Q^* such that $\rho K \subset K$ for all $\rho \geq 0$*

(i) *If K is strictly separating for Q, then $K_{\Psi X}$ is strictly separating for $C\Psi(X,Q)^*$.*

(ii) *If the intersection of K with the polar of each neighborhood $v \in V$ is $w(Q^*,Q)$-closed, then the intersection of $K_{\Psi X}$ with the polar of each neighborhood $v_\psi \in V_\Psi$ is $w(C\Psi(X,Q)^*,C\Psi(X,Q))$-closed.*

Proof. (i): Let $v \in V$ and $\psi \in \Psi$. By assumption on K there exist an equicontinuous subset K_0 of K, i.e. $K_0 \subset w_Q^\circ$ for some $w \in V$, and some $\rho > 0$ such that for all $a,b \in Q$

$$a \not\leq b+v \text{ implies } \mu(a) > \mu(b)+\rho \text{ for some } \mu \in K_0.$$

Now set $K_{X0} = \{\mu_x \mid \mu \in \psi(x)K_0\}$ which is contained in the polar of w_ψ by 1.7(i), whence equicontinuous in $C\Psi(X,Q)^*$. Finally, if $f \not\leq g+v_\psi$, then $\psi(x)f(x) \not\leq \psi(x)g(x)+v$ for some $x \in X$, whence $\psi(x)\mu(f(x)) > \psi(x)\mu(g(x))+\rho$ for some $\mu \in K_0$; thus $\psi(x)\mu_x(f) > \psi(x)\mu_x(g)+1$ with $\psi(x)\mu_x \in K_{X0}$. Hence K_{X0} is strictly separating for $C\Psi(X,Q)$. Moreover, as K is closed with respect to multiplication by positive scalars, K_{X0} is a subset of $K_{\Psi X}$.

In order to verify (ii), let $v_\psi \in V_\Psi$ and let $\Phi \in C\Psi(X,Q)^*$ be in the closure of $K_{\Psi X} \cap v_\psi^\circ$. We have to show that Φ is already contained in $K_{\Psi X} \cap v_\psi^\circ$. This is obvious for $\Phi = 0$; thus we may assume that $\Phi \neq 0$ and consider a net of non-zero functionals $(\mu_{\alpha x_\alpha})_{\alpha \in A}$ in $K_{\Psi X} \cap v_\psi^\circ$ converging to Φ. By Lemma 1.7, for every α we have $x_\alpha \in \operatorname{supp} \psi$ and $\mu_\alpha \in \rho v_Q^\circ$ for all $\rho > \psi(x_\alpha)$. We shall distinguish two cases:

Firstly, assume that $(x_\alpha)_{\alpha \in A}$ converges to infinity; i.e. for every compact subset Y of X there is $\alpha_0 \in A$ such that $x_\alpha \notin Y$ for all $\alpha \geq \alpha_0$. We shall show that $(\mu_{\alpha x_\alpha})_{\alpha \in A}$ converges to $\Phi = 0 \in K_{\Psi X}$ in this case: Let $f \in C\Psi(X,Q)$ and $\varepsilon > 0$. By 1.2 there is a compact subset Y of X such that

$$\psi(x)f(x) \leq \varepsilon v \text{ and } 0 \leq \psi(x)f(x) + \varepsilon v \text{ for all } x \in X \setminus Y.$$

For all $\alpha \geq \alpha_0$ we have $x_\alpha \in X\setminus Y$. Next we recall that these elements x_α are all contained in the support of ψ and may therefore be approximated by elements $y \in X\setminus Y$ such that $\psi(y) > 0$. The function f is v-bounded in these y. But f is continuous with respect to the symmetric topology on Q, and the subcone of v- bounded elements is both open and closed in Q in this topology. Thus its inverse image under f is a closed subset of X. The latter shows that the values of f are v-bounded in the elements x_α, for all $\alpha \geq \alpha_0$, as well. For the corresponding functionals of our subnet we conclude

$$|\mu_{\alpha x_\alpha}(f)| = |\mu_\alpha(f(x_\alpha))| \leq \varepsilon$$

whenever $\psi(x_\alpha) \neq 0$, and $\mu_{\alpha x_\alpha}(f) = \mu_\alpha(f(x_\alpha)) = 0$ if $\psi(x_\alpha) = 0$, thus verifying our claim.

Now, secondly, assume that there is a compact subset Y of X such that for every $\alpha \in A$ there is a $\beta_\alpha \geq \alpha$ such that $x_{\beta_\alpha} \in Y$. Then $(x_\alpha)_{\alpha \in A}$ admits a subnet $(x_\beta)_{\beta \in B}$ converging to some limit point $x \in Y \cap \operatorname{supp} \psi$. As ψ is upper semicontinuous, given a fixed $\rho > \psi(x)$, there is $\beta_0 \in B$ such that $\psi(x_\beta) < \rho$, whence $\mu_\beta \in \rho v_Q^\circ$, for all $\beta \geq \beta_0$. Thus, by our assumption on K we may assume (after selecting a further subnet) that the net $(\mu_\beta)_{\beta \in B}$ is convergent as well with limit point $\mu \in K$ with the additional property that $\mu \in \rho v_Q^\circ$. We shall show that the net $(\mu_{\beta x_\beta})_{\beta \in B}$, converges to μ_x, whence $\mu_x = \Phi$, as the whole net $(\mu_{\alpha x_\alpha})_{\alpha \in A}$ was supposed to converge to Φ:

Let $f \in C\Psi(X,Q)$ and $\varepsilon > 0$. Because of the continuity of f there is a neighborhood U of x in X such that

$$f(y) \leq f(x) + (\varepsilon/2\rho)v \text{ and } f(x) \leq f(y) + (\varepsilon/2\rho)v \text{ for all } x \in U.$$

Thus, we find $\beta_1 \in B$ such that for all $\beta \geq \beta_1$ we have

$$x_\beta \in U,\ \mu_\beta \in \rho v_Q^\circ \text{ and } \mu_\beta(f(x)) \leq \mu(f(x)) + \varepsilon/2.$$

Furthermore, this yields for all such β

$$\mu_{\beta x_\beta}(f) = \mu_\beta(f(x_\beta)) \leq \mu_\beta(f(x)) + \varepsilon/2 \leq \mu(f(x)) + \varepsilon = \mu_x(f) + \varepsilon.$$

To obtain a converse inequality we have to consider two cases: If $\mu_x(f) < +\infty$, then we choose the index $\beta_2 \geq \beta_1$ such that for all $\beta \geq \beta_2$ we have in addition

$$\mu(f(x)) \leq \mu_\beta(f(x)) + \varepsilon/2.$$

But this shows that

$$\mu_x(f) = \mu(f(x)) \leq \mu_\beta(f(x)) + \varepsilon/2 \leq \mu_\beta(f(x_\beta)) + \varepsilon = \mu_{\beta x_\beta}(f) + \varepsilon$$

holds. If, on the other hand, $\mu_x(f) = +\infty$, then we require instead that $1/\varepsilon \leq \mu_\beta(f(x))$ holds for all $\beta \geq \beta_2$, whence

$$1/\varepsilon < \mu_\beta(f(x)) \leq \mu_\beta(f(x_\beta)) + \varepsilon/2 = \mu_{\beta x_\beta}(f) + \varepsilon.$$

Thus, the net $(\mu_{\beta x_\beta})_{\beta \in B}$, converges to μ_x in any case. All left to show in this case is that $\mu_x \in K_{\Psi X}$: As we already remarked above, the preceding argument implies that $\mu_x = \Phi$; thus the functional $\mu \in K$ cannot depend on the special choice of the fixed number $\rho > \psi(x)$. Thus we have $\mu \in \rho v_Q^\circ$ for all $\rho > \psi(x)$, and using 1.7, $\mu_x \in K_{\Psi X}$, indeed. This completes our proof.

For Banach spaces Q and compact sets X the following result is due to Singer [56]. Combining Proposition 1.8 with the observation of Proposition 1.6 from Ch. IV we obtain immediately:

1.9 Proposition. *Suppose that all elements of the locally convex cone* (Q,V) *are bounded. Let* K *be a subset of* Q^* *such that* $\rho K \subset K$ *for all* $\rho \geq 0$. *Suppose that for every* $v \in V$ *the set* $K \cap v_Q^\circ$ *is* $w(Q^*,Q)$*-closed, and for all* $a,b \in Q$

$$a \not\leq b+v \quad \text{implies} \quad \mu(a) > \mu(b)+1 \quad \text{for some } \mu \in K \cap v_Q^\circ .$$

Let X *be a locally compact Hausdorff space and* Ψ *a family of weight functions on* X. *Then for every* $v_\psi \in V_\Psi$, *all non-zero extreme points of its polar are of the type* $\psi(x)\mu_x$ *for some* $x \in X$ *and some extreme point* μ *of* v_Q°.

Proof. By the preceding Proposition 1.8 the set $K_{\Psi X}$ fulfils the assumptions of Proposition IV1.6, thus contains all extreme points of v_ψ°. If $0 < \rho < \psi(x)$ or if μ is non-extreme in v_Q° then it is clear that the functional $\rho\mu_x$ cannot be extreme in v_ψ°. If $\psi(x) = 0$ and $\mu_x \in K_{\Psi X} \cap v_\psi^\circ$, then our characterization of this intersection yields $\mu = 0$, as Q is supposed to contain only bounded elements.

In Chapter II.5 we introduced the concept of (M-)uniformly up- and down-directed locally convex cones with the semiinterpolation property (USIP and DSIP). This yielded a helpful description of the extreme points of the polars of neighborhoods (Theorem II.6.7). As we mentioned in Example II.5.10, in general we cannot expect for a uniformly directed cone (Q,V) that cones of continuous Q-valued functions on a topological space always carry the same property. But in some important special cases this may be guaranteed for Nachbin cones. The following uses the observation we made at that time (c.f Example II.5.10):

1.10 Proposition. *Let* (Q,V) *be an (M-)uniformly down-directed locally convex cone.*

(i) *If* Q *is an* $\wedge$*-semilattice with* DSIP *then* $C\Psi(X,Q)$ *is also an (M-)uniformly down-directed* $\wedge$*-semilattice with* DSIP.

(ii) *If* $-V \subset Q$, *if* X *is compact and if all weight functions are strictly positive and continuous, then* $C\Psi(X,Q)$ *is M-uniformly down-directed. If* Q *is even a vector space then* $C\Psi(X,Q)$ *has the* DSIP.

Similar statements hold for the (M-)uniformly up-directed case. The first requirement in (ii) then is $V \subset Q$.

Proof. (i): For $f,g \in C\Psi(X,Q)$ we know from Example II.5.10 that the function $f \wedge g$ is continuous as well. We still have to verify the condition in 1.2 in order to prove that

$f \wedge g \in C\Psi(X,Q)$: Let $v \in V$ and $\psi \in \Psi$. By Proposition II.5.6 there is $v' \in V$ such that for all $a,b \in Q$

$$0 \le a+v' \text{ and } 0 \le b+v' \text{ implies } 0 \le a \wedge b+v.$$

Now by 1.2 we find a compact subset Y of X such that

$$h(x) \text{ is bounded, } \psi(x)h(x) \le v' \text{ and } 0 \le \psi(x)h(x)+v' \text{ for all } x \in X \setminus Y$$

holds for both $h=f$ and $h=g$. But by the above this shows that

$$(f \wedge g)(x) \text{ is bounded, } \psi(x)(f \wedge g)(x) \le v' \text{ and } 0 \le \psi(x)(f \wedge g)(x)+v' \text{ for all } x \in X \setminus Y$$

holds as well. Finally, as Q has DSIP, the distributive law holds for $\wedge$ in Q. But this induces the same property for Q-valued functions, whence DSIP holds for $C\Psi(X,Q)$ as well.

(ii): Under the assumptions on X and Ψ, all the functions $1/\psi$ for $\psi \in \Psi$ are contained in $C\Psi(X)$. But the functions $(1/\psi)v$ are just a neighborhood base of $C\Psi(X,Q)$. Thus, $-V \subset Q$ implies $-V_\Psi \subset C\Psi(X,Q)$, and (ii) follows from Remarks II.5.2(b) and II.5.8(b).

Now we are in a position to formulate a Choquet-Deny type theorem (c.f. Ch. III, Corollary 3.6) for Nachbin cones of continuous cone-valued functions. The following combines the results of Chapter III with our preceding observations:

1.11 Proposition. *Suppose that all elements of the locally convex cone (Q,V) are bounded. Let K be a subset of Q^* such that $\rho K \subset K$ for all $\rho \ge 0$. Suppose that for every $v \in V$ the set $K \cap v^\circ_Q$ is $w(Q^*,Q)$-closed, and that for all $a,b \in Q$*

$$a \not\le b+v \text{ implies } \mu(a) > \mu(b)+1 \text{ for some } \mu \in K \cap v^\circ_Q.$$

Let X be a locally compact Hausdorff space and Ψ a family of weight functions on X. Let C_0 be an M-uniformly down-directed subcone of $C\Psi(X,Q)$. If the function $f \in C\Psi(X,Q))$ is C_0-superharmonic in all points of $K_{\Psi X}$ [i.e. if for all $x \in X$ such that $\psi(x) \ne 0$ for at least one weight function $\psi \in \Psi$, $\mu \in K$ and all functionals $\phi \in C\Psi(X,Q)^*$

$$\phi(g) \le \mu(g(x)) \text{ for all } g \in C_0 \text{ implies } \phi(f) \le \mu(f(x))\ ],$$

then f is contained in the closure of C_0 with respect to the symmetric topology.

An analogous statement holds for uniformly up-directed subcones of $C\Psi(X,Q)$ and subharmonic functions.

Proof. It follows from Proposition 1.9 that for each $v_\psi \in V_\Psi$ the subset $K_{\Psi X}$ contains all extreme points of v_ψ°. Note that the condition of Proposition 1.9 implies C_0-superharmonicitiy in all points of $K_{\Psi X}$ Thus, we may apply Theorem 3.4 from Ch. III which shows C_0-superharmonicitiy for the function f in all points of v_ψ°, whence in all of $C\Psi(X,Q)^*$. Corollary 1.5 from Ch. III then yields our claim.

2. A criterion for super- and subharmonicity.

Throughout this section let X be a locally compact Hausdorff space, Ψ a family of weights on X and (Q,V) a locally convex cone. Let C_0 be a subcone of $C\Psi(X,Q)$. The condition for a cone-valued function $f \in C\Psi(X,Q)$ to be C_0-superharmonic in the elements of $K_{\Psi X}$ will not be easy to check as it involves arbitrary functionals in the dual cone of $C\Psi(X,Q)$ which may be hard to handle. As a remedy we offer a technical but useful criterion for super- and subharmonicity which only involves the set $K_{\Psi X}$ itself. Firstly, we return to the real-valued case, where the elements of the dual cone may be represented as measures. As usual, we endow $\mathbf{R}$ with the neighborhoods $V = \{\rho > 0\}$. The following is a generalization of Proposition IV.2.2. It contains in particular the result by Bauer and Donner [11] for spaces $C_0(X)$:

2.1 Proposition. *Let X be a locally compact Hausdorff space and Ψ be a family of weights on X. Let G be a subspace of $C\Psi(X)$. For a function $f \in C\Psi(X)$ the following conditions are equivalent:*

(i) *For every net $(T_\alpha)_{\alpha \in A}$ of equicontinuous positive linear operators on $C\Psi(X)$:*
$$T_\alpha(g) \to g \text{ for all } g \in G, \text{ implies } T_\alpha(f) \to f.$$

(ii) *For every $x \in X$ such that $\psi(x) \neq 0$ for some $\psi \in \Psi$ we have*
$$f(x) = \sup_{\substack{\varepsilon>0 \\ \psi \in \Psi}} \inf \{g(x) \mid g \in G,\ \psi f \leq \psi g + \varepsilon\}$$
$$= \inf_{\substack{\varepsilon>0 \\ \psi \in \Psi}} \sup \{g(x) \mid g \in G,\ \psi g \leq \psi f + \varepsilon\}.$$

(iii) *For every $x \in X$ such that $\psi(x) \neq 0$ for some $\psi \in \Psi$ and for every bounded positive Radon measure μ on X and every weight function $\psi \in \Psi$:*
$$\mu(\psi g) = g(x) \text{ for all } g \in G, \text{ implies } \mu(\psi f) = f(x).$$

Convergence in (i) is meant with respect to the symmetric topology of $C\Psi(X)$.

Proof. As we remarked in 1.5, the dual cone of $C\Psi(X)$ consists exactly of the measures $\psi\mu$, where μ is a bounded positive Radon measure on X. Using Proposition 1.8 with $K = \{\rho \geq 0\}$, the Convergence Theorem IV.1.11 yields the implication (iii) $\Rightarrow$ (i). (ii) and (iii) are equivalent by the Sup-Inf-Theorem III.1.3, and the implication (i) $\Rightarrow$ (iii) follows with the same construction as in Proposition IV.2.2: The neighborhood U of x may be chosen relatively compact which guarantees that the resulting operator T_U indeed maps $C\Psi(X)$ into $C\Psi(X)$.

2.2 Korovkin systems for $C\Psi(X)$. As usual, a subset M of $C\Psi(X)$ is said to be a *Korovkin system for $C\Psi(X)$* if every function in $C\Psi(X)$ fulfils the equivalent conditions of Proposition 2.1, where G is meant to be the linear span of M. For our purposes, however, dealing with cones rather than with linear spaces we have to complement this definition: We shall say that the subset M of $C\Psi(X)$ is a *$^+$Korovkin system for $C\Psi(X)$* if every positive function in $C\Psi(X)$ is G_0-superharmonic in all elements of $K_{\Psi X}$, where now G_0 is meant to be the subcone of $C\Psi(X)$ generated by M; more precisely, if the following condition holds:

For every $x \in X$ *such that* $\psi(x) \neq 0$ *for at least one* $\psi \in \Psi$ *and for every bounded positive Radon measure* μ *on* X *and every weight function* $\psi \in \Psi$

$$\mu(\psi g) \leq g(x) \text{ for all } g \in M, \text{ implies } \psi\mu = \lambda\varepsilon_x \text{ for some } 0 \leq \lambda \leq 1.$$

By the Sup-Inf-Theorem II.1.3 this condition is equivalent to:

For every $x \in X$ *such that* $\psi(x) \neq 0$ *for at least one* $\psi \in \Psi$ *we have*

$$f(x) = \sup_{\substack{\varepsilon > 0 \\ \psi \in \Psi}} \inf \{g(x) \mid g \in G_0,\ \psi f \leq \psi g + \varepsilon\} \text{ for all } f \in C\Psi(X)^+.$$

(By $C\Psi(X)^+$, as usual we denote the positive cone in $C\Psi(X)$. Furthermore, recall that for a positive Radon measure ν and a point $x \in X$, $\nu(f) \leq f(x)$ for all $f \in C\Psi(X)^+$ means that $\nu = \lambda\varepsilon_x$ for some $0 \leq \lambda \leq 1$.)

2.3 Remarks and examples. (a) If for each $x \in X$ the $^+$Korovkin system M contains a function g_x such that $g_x(x) < 0$, then the first version of our condition shows that $(\psi\mu)(g) \leq g(x)$ for all $g \in M$ implies $\psi\mu = \lambda\varepsilon_x$ for some $0 \leq \lambda \leq 1$, and $(\psi\mu)(g_x) = \lambda g_x(x) \leq g_x(x)$ in particular, whence even $\lambda = 1$. Thus every function in $C\Psi(X)$ is G_0-superharmonic in all point evaluations in X, and M is a Korovkin system for $C\Psi(X)$ in the usual sense. On the other hand, if M is a Korovkin system for $C\Psi(X)$, then obviously M-M is a $^+$Korovkin system for $C\Psi(X)$.

(b) For $X = [0,1]$, $\Psi = \{1\}$, i.e. $C\Psi(X) = C[0,1]$ with the usual sup-norm topology, the functions $f_0(x) = 1$, $f_1(x) = -x$, and $f_2(x) = x^2$ form a $^+$Korovkin system in our sense as the cone generated by them contains all the functions $f(x) = (x-x_0)^2$, $x_0 \in [0,1]$, which, together with the positive constants, guarantee our condition.

(c) For $X = \mathbf{R}$, $\Psi = \{1\}$, i.e. $C\Psi(X) = C_0(\mathbf{R})$ with the sup-norm topology, the functions $f_0(x) = e^{-x^2}$, $f_1(x) = -xe^{-x^2}$, and $f_2(x) = x^2e^{-x^2}$ form a $^+$Korovkin system.

(d) For $X = \mathbf{R}$ and Ψ consisting of the characteristic functions of the compact subsets of $\mathbf{R}$, i.e $C\Psi(X) = C(\mathbf{R})$ with the topology of compact convergence, the same functions as in (b) form a $^+$Korovkin system.

Using this we formulate our criterion. We shall first give a simplified version which avoids some of the technical complications of the general result. For some applications, however, it will already prove sufficient. To prepare our result, note that for a positive function $f \in C\Psi(X)$ and a bounded element $a \in \mathbf{Q}$, the $\mathbf{Q}$-valued function $f{\cdot}a$, i.e. the mapping

$$x \to f(x)a \ : \ X \to \mathbf{Q},$$

is contained in the Nachbin cone $C\Psi(X,\mathbf{Q})$. Moreover, $(-f){\cdot}a = f{\cdot}(-a)$ may be added to $C\Psi(X,\mathbf{Q})$. For unbounded elements $a \in \mathbf{Q}$, however, only positive functions f may be considered, and either f or all weight functions need to have a compact support (c.f. the definition of a Nachbin cone in 1.2).

2.4 Theorem. *Let the locally convex cone (Q,V) be a vector space. Let X be a locally compact Hausdorff space, Ψ a family of weight functions on X, M a ${}^{+}$Korovkin system for $C\Psi(X)$, and C_0 a subcone of $C\Psi(X,Q)$ such that the following holds:*

(i) *For every $a \in Q$ and $v \in V$ there is an element $0 \le e \in Q$ such that $a \le e+v$, and all the functions*

$$x \to g(x)e \ :\ X \to Q, \quad g \in M$$

are contained in C_0.

(ii) *For every $a \in Q$ there is a strictly positive function $p \in C\Psi(X)$ such that the function*

$$x \to p(x)a \ :\ X \to Q$$

is contained in C_0.

Then every function $f \in C\Psi(X,Q)$ is C_0-superharmonic on $Q^*_{\Psi X}$.

Theorem 2.4 is seen to be an immediate corollary of the following more general result: Set $K = Q^*$ in 2.5 and recall that Q is a locally convex ordered topological vector space in 2.4, Q^* the positive cone in its dual. Conditions (i) and (ii) in 2.4 obviously imply (i) and (ii) in 2.5 for all $\mu \in Q^*$.

2.5 Theorem. *Let (Q,V) be a locally convex cone. Let K be a subset of Q^* such that $\rho K \subset K$ for all $\rho \ge 0$. Suppose that K is strictly separating for Q and that its intersection with the polar of each neighborhood is $w(Q^*,Q)$-closed. Let X be a locally compact Hausdorff space, Ψ a family of weight functions on X, M a ${}^{+}$Korovkin system for $C\Psi(X)$, and C_0 a subcone of $C\Psi(X,Q)$. Suppose that for the functional $\mu \in K$ the following holds:*

(i) *There is an element $0 \le e_\mu \in Q$ such that $0 < \mu(e_\mu) < +\infty$, and all the functions*

$$x \to g(x)e_\mu \ :\ X \to Q, \quad g \in M$$

are contained in C_0.

(ii) *For every $\tau \in K$ which neither is a multiple of μ nor vanishes on all bounded elements of Q, there is an element $0 \le c_{\mu\tau} \in Q$ such that $0 = \mu(c_{\mu\tau}) < \tau(c_{\mu\tau})$, and a strictly positive function $p_{\mu\tau} \in C\Psi(X)$ such that the function*

$$x \to p_{\mu\tau}(x)c_{\mu\tau} \ :\ X \to Q$$

is contained in C_0.

Let $x \in X$ be such that $\psi(x) \ne 0$ for at least one weight function $\psi \in \Psi$. If C_0 contains some bounded function g such that $\mu(g(x)) > 0$ (resp. $\mu(g(x)) < 0$), then every bounded function $f \in C\Psi(X,Q)$ such that $\mu(f(x)) \ge 0$ (resp. $\mu(f(x)) \le 0$) is C_0-superharmonic in μ_x.

Proof. We shall use arguments from measure theory to prove this result:
Let $\mu \in K$ be as in the theorem, $x \in X$ such that $\psi(x) \neq 0$ for some $\psi \in \Psi$. Let $\pi \in C\Psi(X,Q)^*$ be any functional such that $\pi(g) \leq \mu_x(g)$ for all $g \in C_0$. We have to verify that this implies $\pi(f) \leq \mu_x(f)$ for those functions $f \in C\Psi(X,Q)$ as in the theorem:
Let $v \in V$ and $\phi \in \Psi$ be such that both functionals μ_x and π are contained in v_ϕ°. By our assumption on K there is a $w(Q^*,Q)$-compact equicontinuous subset K_0 of K and some $\rho > 0$ such that for $a,b \in Q$, $a \not\leq b+v$ implies $\tau(a) > \tau(b)+\rho$ for some $\tau \in K_0$. We may assume that K_0 contains μ. Thus, for any two functions $f,g \in C\Psi(X,Q)$ such that $f \not\leq g+v_\Phi$ we have $\phi(y)f(y) \not\leq \phi(y)g(y)+v$ for some $y \in X$, whence

$$\phi(y)\tau_y(f) > \phi(y)\tau_y(g)+\rho \quad \text{for some } y \in X,\ \tau \in K_0.$$

Next we consider the locally compact space $X \times K_0$ and the weight functions ψ on $X \times K_0$ associated with the weight functions $\psi \in \Psi$ and defined by $\psi(y,\tau) = \psi(y)$. Let $\overline{C\Psi}(X \times K_0)$ be the locally convex cone of continuous $\overline{R}$-valued functions on $X \times K_0$, where $\overline{R}$ is endowed with its usual topology and the neighborhoods are defined via the weight functions ψ similar to a Nachbin cone. Note, however, that $\overline{C\Psi}(X \times K_0)$ is not a Nachbin cone in our sense, as $\overline{R}$ with this topology is not a locally convex cone (the symmetric cone topology would isolate $+\infty$). The subcone $C\Psi(X \times K_0)$ of real valued functions, however, is a Nachbin cone. There is a (not necessarily one-to-one) mapping from $C\Psi(X,Q)$ into $\overline{C\Psi}(X \times K_0)$:

$$f \to F \quad \text{where } F \text{ is the function on } X \times K_0 \text{ defined by } F(y,\tau) = \tau_y(f) = \tau(f(y)).$$

Note that bounded functions in $C\Psi(X,Q)$ are mapped into $C\Psi(X \times K_0)$. By the above $\phi F \leq \phi G+1$ implies $f \leq g+(1/\rho)v_\phi$, whence $\pi(f) \leq \pi(g)+1/\rho$. Thus π may be considered as a u-continuous linear functional on a subcone of $\overline{C\Psi}(X \times K_0)$ and, by Ch. II.2.9, may be extended to a u-continuous linear functional $\tilde{\pi}$ on all of $\overline{C\Psi}(X \times K_0)$; more precisely: The extension $\tilde{\pi}$ may be chosen such that

$$\phi F \leq \phi G+1 \quad \text{for arbitrary } F,G \in \overline{C\Psi}(X \times K_0) \text{ implies } \tilde{\pi}(F) \leq \tilde{\pi}(G)+1/\rho.$$

By 1.5, the restriction of $\tilde{\pi}$ to $C\Psi(X \times K_0)$ coincides with $\phi\Pi$ for some positive bounded Radon measure Π on $X \times K_0$. We shall proceed to investigate the support of $\phi\Pi$:

Firstly, let $\tau \in K_0$ be a functional as in (ii), and choose the function $p_{\mu\tau} \in C\Psi(X)$ and the element $c_{\mu\tau} \in Q$ be as in that condition. By $p \in C_0$ we denote the mapping $y \to p_{\mu\tau}(y)c_{\mu\tau}$ and by P, as before, its image as a function in $\overline{C\Psi}(X \times K_0)$, i.e.

$$P(y,\sigma) = p_{\mu\tau}(y)\,\sigma(c_{\mu\tau}) \quad \text{for } (y,\sigma) \in X \times K_0.$$

Note that P is not necessarily contained in the subcone $C\Psi(X \times K_0)$ as $c_{\mu\tau}$ was not required to be bounded, thus P may attain infinite values. But from our assumption on the functional π we know

$$\tilde{\pi}(P) = \pi(p) \leq \mu_x(p) = p_{\mu\tau}(x)\mu(c_{\mu\tau}) = 0.$$

Setting $Q = P \wedge 1 \in C\Psi(X \times K_0)$ and using the monotony of $\tilde{\pi}$, this shows

$$\phi\Pi(Q) = \tilde{\pi}(Q) \leq \tilde{\pi}(P) \leq 0.$$

Finally, Q is a positive function, and as $p_{\mu\tau}(y) > 0$ for all $y \in X$, we have $Q(y,\tau) > 0$. Thus, for all $y \in X$ and every $\tau \in K$ which is neither a multiple of μ nor vanishes on all bounded elements of Q, we infer that (y,τ) cannot be contained in the support of $\phi\Pi$.

Now, secondly, let the element $e_\mu \in Q$ be as in condition (i). We may assume that $\mu(e_\mu) = 1$. Let $x \neq y \in X$ and $(y,\tau) \in X \times K_0$ such that $\tau = \alpha\mu$ for some $0 \neq \alpha \in \boldsymbol{R}$. We shall exclude this point from the support of $\phi\Pi$ as well: From $e_\mu \geq 0$ we conclude, as $\tau \in K_0 \subset Q^*$, that $\tau(e_\mu) = \alpha\mu(e_\mu) = \alpha > 0$. Now choose a positive function $f \in C\Psi(X)$ such that $0 = f(x) < f(y)$. Given $\varepsilon > 0$, by 2.2(i) we find a function g in the subcone G_0 of $C\Psi(X)$ generated by the ${}^+$Korovkin system M such that

$$g(x) \leq f(x) + \varepsilon = \varepsilon \text{ and } \phi f \leq \phi g + \varepsilon.$$

Now we introduce the following functions in $\overline{C}\Psi(X \times K_0)$:

$$F(z,\sigma) = f(z)(\sigma(e_\mu) \wedge 1), \; H(z,\sigma) = g(z)(\sigma(e_\mu) \wedge 1) \text{ and } G(z,\sigma) = g(z)\sigma(e_\mu),$$

for $(z,\sigma) \in X \times K_0$. The functions F and H are bounded on $X \times K_0$ and contained in $C\Psi(X \times K_0)$, thus the above inequalities imply

$$\phi(z)F(z,\sigma) \leq \phi(z)H(z,\sigma) + \varepsilon \quad \text{for all } (z,\sigma) \in X \times K_0,$$

whence

$$\phi\Pi(F) = \tilde{\pi}(F) \leq \tilde{\pi}(H) + \varepsilon/\rho$$

by the way we constructed the functional $\tilde{\pi}$. Using the monotonicity of $\tilde{\pi}$ and $g \cdot e_\mu \in \boldsymbol{C_0}$ we continue

$$\tilde{\pi}(H) \leq \tilde{\pi}(G) = \pi(g \cdot e_\mu) \leq \mu_x(g \cdot e_\mu) \leq \varepsilon\mu(e_\mu) = \varepsilon.$$

Combined, this yields $\phi\Pi(F) \leq \varepsilon(1 + 1/\rho)$ and even $\phi\Pi(F) \leq 0$, as $\varepsilon > 0$ was selected arbitrarily. The function F is positive on $X \times K_0$, and as $F(y,\tau) = F(y,\alpha\mu) = f(y)(\alpha\mu(e_\mu) \wedge 1) = f(y)(\alpha \wedge 1) > 0$, this excludes (y,τ) form the support of $\phi\Pi$, indeed.

Summarizing, we conclude that the support of $\phi\Pi$ contains only points of $X \times K_0$ of two different categories: (y,τ), where $y \in X$ is arbitrary and $\tau \in K_0$ vanishes on all bounded elements of Q, and $(x,\alpha\mu)$ for $\alpha \geq 0$ and the given elements $x \in X$ and $\mu \in K_0$. Now a simple argument shows that there is some $\lambda \geq 0$ such that for every bounded function $f \in C\Psi(X,Q)$ and its representation F as a function in $C\Psi(X \times K_0)$ we have

$$\pi(f) = \tilde{\pi}(F) = \phi\Pi(F) = \lambda\mu(f(x)).$$

Finally, if there is a bounded function $g \in \boldsymbol{C_0}$ such that $\mu(g(x)) > 0$ (resp. $\mu(g(x)) < 0$) we infer $\pi(g) = \lambda\mu(g(x)) \leq \mu(g(x))$, whence $\lambda \leq 1$ (resp. $\lambda \geq 1$). But this shows that $\pi(f) = \lambda\mu(f(x)) \leq \mu(f(x)) = \mu_x(f)$ for all $f \in C\Psi(X,Q)$ such that $\mu(f(x)) \geq 0$ (resp. $\mu(f(x)) \leq 0$), thus completing our proof.

2.6 Remarks. (a) The criterion of Theorem 2.5 applies to locally convex cones in general: Every locally convex cone (Q,V) may be identified with $C\Psi(X,Q)$, where X is a singleton set, $\Psi = \{1\}$, with the Korovkin system M consisting of the one-function.

(b) A similar criterion might be formulated for subharmonicity using the functions $f \in -M$ in (i) and negative elements $c_{\mu\tau} \in Q$ in (ii). But then all functions involved in the corresponding criterion need to be bounded; thus we may well consider their negatives and superharmonicity with respect to them instead.

(c) Note that the combination of Theorem 2.5 with Proposition 1.11 yields a Stone-Weierstraß type theorem for $C\Psi(X,Q)$.

(d) Note that 2.4 is a somewhat crude simplification of 2.5, as it requires the subcone C_0 to contain *all* "constant" functions (typically, we think of the functions $p_{\mu\tau}$ as of replacements for the constant function 1). This will become clear in the following section when we shall deal with different versions of Korovkin type theorems for set-valued functions. The reader may also compare Theorem 2.4 to Prolla's work in [44].

2.7 Example: Approximation for stochastic processes. The following application is due to M. Weba [60]: In statistics, it frequently occurs that a real-valued function is not known precisely. So its value for a given argument x in a compact Hausdorff space X may be considered as random function $\phi(\omega)$ depending on elements of an underlying probability space (Ω,A,σ). Those functions should be continuous with respect to the compact topology of X and the L^p-integration norm for the random functions. Positivity is considered with respect to the usual order of L^p; i.e. we are dealing with $C(X,L^p)$, $1 \leq p < +\infty$, endowed with the topology of uniform convergence. The measure σ is supposed to be a probability measure. Thus, L^p contains the constant functions, and $C(X)$ may be identified with the subspace of functions in $C(X,L^p)$ not depending on ω, i.e. the "sure" functions. The linear operators under consideration should not only be positive but satisfy two additional properties. For $f \in C(X)$ and $\phi \in L^p$ we denote by $f\cdot\phi \in C(X,L^p)$ the function $x \to f(x)\phi$. A linear operator T on $C(X,L^p)$ is called

(i) *weakly E-commutative* if it maps $C(X)$ into $C(X)$.

(ii) *stochastically simple* if for all $f \in C(X)$ and $\phi \in L^p$ it maps $f\cdot\phi$ into $T(f)\cdot\phi$.

By direct methods Weba proves the following Korovkin type theorem for stochastic processes on $X = [0,1]$:

Let $(T_\alpha)_{\alpha\in A}$ be an equicontinuous net of positive linear weakly E-commutative and stochastically simple operators on $C([0,1],L^p)$. Then $T_\alpha(g) \to g$ for the sure functions $g(x) = 1, x, x^2$ implies $T_\alpha(f) \to f$ for all $f \in C_s([0,1],L^p)$.

We may recover and considerably generalize this results using the simplified version 2.4 of our criterion: Let $Q = L^p$, the ordered vector space as a locally convex cone, endowed with its usual neighborhood system as in Example I.2.7. Let X be a locally compact space and Ψ a family of weight functions on X, and let M be a Korovkin system for $C\Psi(X)$ containig a strictly positive function p. Considering Theorem 2.4, we shall operate with the $^+$Korovkin system M-M. For $a \in A$ by χ_a we denote its characteristic function, i.e.

$$\chi_a(\omega) = \begin{cases} 1, & \text{if } \omega \in a, \\ 0, & \text{if } \omega \notin a. \end{cases}$$

Again, $C\Psi(X)$ is identified with the subspace of "sure" functions in $C\Psi(X,L^p)$. In an obvious generalization of the above notations we shall say that a linear operator T on $C\Psi(X,L^p)$

is *weakly E-commutative* if it maps $C\Psi(X)$ into $C\Psi(X)$, and *stochastically simple* if for all $f \in C\Psi(X)$ and $\phi \in L^p$ it maps $f\cdot\phi$ into $T(f)\cdot\phi$. With these definitions it is obvious that for every equicontinuous net $(T_\alpha)_{\alpha\in A}$ of positive linear weakly E-commutative and stochastically simple operators on $C\Psi(X,L^p)$, $T_\alpha(p) \to p$ implies $T_\alpha(p\cdot\chi_a) \to p\cdot\chi_a$ for all $a \in A$ and the fixed strictly positive function $p \in M$. As the characteristic functions χ_a are total in L^p we even conclude that $T_\alpha(p\cdot\phi) \to p\cdot\phi$ for all $\phi \in L^p$. Thus, we consider the subspace C_0 of $C\Psi(X,L^p)$ generated by the elements of M (which are sure functions) and the functions $p\cdot\phi$ for all $\phi \in L^p$. Then we immediate check our criterion in 2.4: In (i) we set $e = 1 \in L^p$, (ii) was established above. Thus, every function $f \in C\Psi(X,L^p)$ is seen to be C_0-superharmonic on $Q^*_{\Psi X}$. The Convergence Theorem IV.1.11 then yields Weba's result. We shall formulate this as a corollary:

2.8 Corollary. *Let X be a locally compact Hausdorff space, Ψ a family of weight functions on X, M a Korovkin system for $C\Psi(X)$ which contains a strictly positive function. Let (Ω,A,σ) be a probability space.*
Let $(T_\alpha)_{\alpha\in A}$ be an equicontinuous net of positive linear weakly E-commutative and stochastically simple operators on $C\Psi(X,L^p)$. Then

$T_\alpha(g) \to g$ for the sure functions $g \in M$ implies $T_\alpha(f) \to f$ for all $f \in C\Psi(X,L^p)$.
Convergence is meant with respect to the symmetric topology of $C\Psi(X,L^p)$.

Note that, in fact, we did not use the full strength of all the assumptions on the operators T_α.

2.9 Example: Unbounded interval-valued functions. In the following section we are going to study in detail approximation results for functions whose values are compact convex subsets of a given locally convex cone. Compactness guarantees boundedness, and then we have to apply our preceding criteria only to bounded functions, which significantly reduces the technical efforts. Theorem 2.5, however, does not require the elements of the subcone to be bounded and, thus, applies to more general situations as well. We shall illustrate this with a simple example:
Let $Q = \overline{Conv(R)}$ be the locally convex cone of the non-empty (bounded or unbounded) closed intervals in R, ordered by inclusion and endowed with its canonical neighborhoods. First we identify a suitable subset K of its dual cone:
For $\rho > 0$ define the functionals μ_ρ and ν_ρ on Q by

$$\mu_\rho([\alpha,\beta]) = \rho\beta, \quad \mu_\rho((-\infty,\beta]) = \rho\beta, \quad \mu_\rho([\alpha,+\infty)) = +\infty \quad \text{and} \quad \mu_\rho((-\infty,+\infty)) = +\infty;$$
$$\nu_\rho([\alpha,\beta]) = -\rho\alpha, \quad \nu_\rho((-\infty,\beta]) = +\infty, \quad \nu_\rho([\alpha,+\infty)) = -\rho\alpha \quad \text{and} \quad \nu_\rho((-\infty,+\infty)) = +\infty.$$

The set of all such functionals strictly separating for Q and is closed for multiplication by positive scalars. Its intersection with the polar of each neighborhood, however, is not $w(Q^*,Q)$- closed. In order to remedy this we have to add to "zero-type" functionals 0_μ and 0_ν which are limit points of the sequences $(\mu_{1/n})_{n\in N}$ and $(\nu_{1/n})_{n\in N}$, respectively, and given by

$$0_\mu([\alpha,\beta]) = 0, \quad 0_\mu((-\infty,\beta]) = 0, \quad 0_\mu([\alpha,+\infty)) = +\infty \quad \text{and } 0_\mu((-\infty,+\infty)) = +\infty;$$
$$0_\nu([\alpha,\beta]) = 0, \quad 0_\nu((-\infty,\beta]) = +\infty, \quad 0_\nu([\alpha,+\infty)) = 0 \quad \text{and } 0_\nu((-\infty,+\infty)) = +\infty.$$

These functionals establish the subset K of Q^* which we use in Theorem 2.5 in order to investigate Q-valued functions. As we want to involve constant functions on X, whose values are unbounded elements of Q, we need to require that all weight functions $\psi \in \Psi$ have a compact support (c.f. the definition of Nachbin cones in 1.2). Now consider the positive elements

$$e_\mu = (-\infty, 1], \quad e_\nu = [-1,+\infty), \quad c_{\mu\nu} = (-\infty, 0] \quad \text{and} \quad c_{\nu\mu} = [0,+\infty)$$

in Q which clearly fulfil the requirements (i) and (ii) of Theorem 2.5. (Note that the elements $\tau = 0_\mu$ and $\tau = 0_\nu$ of Q^* vanish on all bounded elements of Q, whence need not be considered for the condition in (ii).) Furthermore, $C\Psi(X)$ contains the constant functions, and we may assume that for $e = [-1,1]$, the function $x \to e$ and its (formal) negative are both contained in the subcone C_0. (More precisely: For the application of Theorem 2.5 to this situation we may extend $\overline{Conv(R)}$ by the formal negatives of its bounded elements; i.e. we consider the cone $\overline{Conv(R)}\text{-}\overline{BConv(R)}$, the latter symbol standing for the bounded intervals in R, which by I.4.8 is a locally convex cone as well. Linear operators on $C\Psi(X,\overline{Conv(R)})$ may be canonically extended to $C\Psi(X,\overline{Conv(R)}\text{-}\overline{BConv(R)})$ (c.f. II.1.5). For a bounded function $f \in C\Psi(X,\overline{Conv(R)})$ and a net $(T_\alpha)_{\alpha\in A}$ of u-continuous operators, convergence in the upper topology, i.e. $T_\alpha(-f) \uparrow -f$, on the formal negative $-f$ of f is equivalent to $T_\alpha(f) \downarrow f$, i.e. convergence in the lower topology on the function f itself. Thus, $T_\alpha(f) \uparrow f$ and $T_\alpha(-f) \uparrow -f$ combined mean $T_\alpha(f) \to f$, i.e. convergence in the symmetric topology, a condition which is formulated in the cone $C\Psi(X,\overline{Conv(R)}$ alone. We shall develop this technique of using the formal negatives of bounded elements with Theorem 2.5 while avoiding them in subsequent convergence results more systematically in the following section.) Then the last requirement of 2.5 is satisfied as well. Given a suitable +Korovkin system of real valued functions, this will show C_0-superharmonicity in the elements $(\mu_\rho)_x$ and $(\nu_\rho)_x$ for all bounded Q-valued functions (and their formal negatives as well). But as we clearly have $(\mu_\rho)_x = (\mu_\rho)_x + (0_\mu)_x$ and $(\nu_\rho)_x = (\nu_\rho)_x + (0_\nu)_x$, the latter also implies C_0-superharmonicity in $(0_\mu)_x$ and $(0_\nu)_x$. (Ch.III, Remark 1.2(a)). We shall summarize this in a corollary. As usual, we identify constant functions with their values.

2.10 Corollary. *Let X be a locally compact Hausdorff space, Ψ a family of weight functions on X such that all $\psi \in \Psi$ have a compact support. Let M be a +Korovkin system of non-negative functions for $C\Psi(X)$, and let the intervals e, e_μ, e_ν, $c_{\mu\nu}$ and $c_{\nu\mu}$ in R be as above.*
Let $(T_\alpha)_{\alpha\in A}$ be a u-equicontinuous net of positive linear operators on $C\Psi(X,\overline{Conv(R)})$. If

$$T_\alpha(h) \uparrow h \quad \text{for all } h \in \overline{M} = \{g\cdot e_\mu \mid g \in M\} \cup \{g\cdot e_\nu \mid g \in M\} \cup \{c_{\mu\nu}, c_{\nu\mu}\}$$

and $$T_\alpha(e) \to e,$$

then $$T_\alpha(f) \to f \text{ for all bounded functions } f \in C\Psi(X,\overline{Conv(R)});$$

i.e. if for all $g \in M$, $\psi \in \Psi$ *and* $\varepsilon > 0$ *there is some* $\alpha_0 \in A$ *such that for all* $\alpha \geq \alpha_0$ *and* $x \in X$

$$\psi(x)T_\alpha(g\cdot e_\mu)(x) \leq \psi(x)(g\cdot e_\mu)(x)+\varepsilon e, \qquad \psi(x)T_\alpha(g\cdot e_\nu)(x) \leq \psi(x)(g\cdot e_\nu)(x)+\varepsilon e,$$
$$\psi(x)T_\alpha(c_{\mu\nu})(x) \leq \psi(x)c_{\mu\nu}+\varepsilon e, \qquad \psi(x)T_\alpha(c_{\nu\mu})(x) \leq \psi(x)c_{\nu\mu}+\varepsilon e,$$
$$\psi(x)T_\alpha(e)(x) \leq \psi(x)e+\varepsilon e \quad \text{and} \quad \psi(x)e \leq \psi(x)T_\alpha(e)(x)+\varepsilon e,$$

then for all bounded functions $f \in C\Psi(X,\overline{Conv(R)})$ *and all* $\varepsilon > 0$ *there is some* $\alpha_0 \in A$ *such that for all* $\alpha \geq \alpha_0$ *and* $x \in X$

$$\psi(x)T_\alpha(f)(x) \leq \psi(x)f(x)+\varepsilon e, \quad \text{and} \quad \psi(x)f(x) \leq \psi(x)T_\alpha(f)(x)+\varepsilon e.$$

Let us reformulate a simplified version of this last corollary for the special case of a compact space X and the usual topology of uniform convergence. Furthermore, we replace nets by sequences of operators. Recall that $T_n(e) \to e$ already implies u-equicontinuity for the sequence $(T_n)_{n\in N}$ of positive linear operators on $C(X,\overline{Conv(R)})$:

2.11 Corollary. *Let* X *be a compact Hausdorff space,* M *a* $^+$*Korovkin system of non-negative functions for* $C(X)$, *and let the intervals* $e, e_\mu, e_\nu, c_{\mu\nu}$ *and* $c_{\nu\mu}$ *in* R *be as above. Let* $(T_n)_{n\in N}$ *be a sequence of positive linear operators on* $C(X,\overline{Conv(R)})$. *If*

$$T_n(h) \uparrow h \quad \text{for all} \quad h \in \overline{M} = \{g\cdot e_\mu \mid g \in M\} \cup \{g\cdot e_\nu \mid g \in M\} \cup \{c_{\mu\nu}, c_{\nu\mu}\}$$

and
$$T_n(e) \to e,$$
then
$$T_n(f) \to f \text{ for all bounded functions } f \in C(X,\overline{Conv(R)}).$$

3. Set-valued functions.

We shall apply the results of the previous sections to study continuous set-valued functions.

Throughout this section we shall assume that the locally convex cone (Q,V) is separated (c.f. I.3.9) and that Q is a vector space. Thus, we are basically in the situation of a locally convex ordered vector space, though our one-sided cone topologies allow some more flexibility. The symmetric topologies, however, were seen to be Hausdorff locally convex ordered vector space topologies.

We shall consider the collection $CConv(Q)$ of all non-empty decreasing subsets A of Q which are compact with respect to the upper topology. It is not difficult to check that such a set A is closed for the lower, hence also the symmetric topology. (But not every lower closed set is upper compact.) As addition and multiplication by a scalar are u-continuous, sums and multiples of sets in $CConv(Q)$ are also compact for the upper topology. Hence, $CConv(Q)$ is a locally convex cone as a subcone of $(\overline{DConv(Q)},\overline{V})$ of all non-empty lower closed convex subsets of Q, as introduced in Ch. I.2.8. Theorem 2.5 will allow us to formulate Korovkin type theorems for Nachbin cones of $CConv(Q)$-valued functions.

We shall also consider the larger cone $\overline{\mathbf{BConv}(Q)}$ of all non-empty bounded decreasing and lower closed convex subsets of Q, which as a subcone of $(\overline{\mathbf{DConv}(Q)},\overline{V})$ is also a locally convex cone. However, there is no longer a sufficient description of the dual cone available in this case, and in order to obtain a Korovkin type theorem for $\overline{\mathbf{BConv}(Q)}$-valued functions we have to resort to our less complete Theorem 2.4.

Let X be a locally compact Hausdorff space and Ψ a family of weights on X. We shall proceed to investigate the Nachbin cone $C\Psi(X,\mathbf{CConv}(Q))$: Clearly, the convex hull, i.e. the supremum in $\overline{\mathbf{DConv}(Q)}$ of two sets in $\mathbf{CConv}(Q)$ is contained in $\mathbf{CConv}(Q)$ as well. Thus, from Proposition 1.10 we know that $C\Psi(X,\mathbf{CConv}(Q))$ is an M-uniformly up-directed $\vee$-semilattice with USIP. In order to apply the results of Sections 1 and 2, in particular Theorem 2.5, we need to identify a suitable subset K of $\mathbf{CConv}(Q)^*$. Recall that there is a natural embedding of Q^* into $\mathbf{CConv}(Q)^*$:

$$\mu \to \overline{\mu} \; : \; Q^* \to \mathbf{CConv}(Q)^*\,,$$

defined by $\overline{\mu}(A) = \sup\{\mu(a) \mid a \in A\}$ for $A \in \mathbf{CConv}(Q)$. Moreover, for every $v \in V$ the restriction of this embedding to v_Q° is even continuous with respect to the topologies $w(Q^*,Q)$ and $w(\mathbf{CConv}(Q)^*,\mathbf{CConv}(Q))$: Let $(\mu_\alpha)_{\alpha \in A}$ be a convergent net in v_Q° with limit μ. For $A \in \mathbf{CConv}(Q)$ and $\varepsilon > 0$ then there are elements $a_1,a_2,\ldots,a_n \in A$ such that the upper neighborhoods $(\varepsilon v)(a_1), (\varepsilon v)(a_2),\ldots,(\varepsilon v)(a_n)$ cover all of A. But this shows that the continuous function on v_E°, $\tau \to (\tau(a_1)\vee\ldots\vee\tau(a_n))$, approximates $\tau \to \overline{\tau}(A)$ by ε. Thus, we have $\overline{\mu}_\alpha(A) \to \overline{\mu}(A)$ as well. The continuity of this embedding is in fact the reason why we have to restrict ourselves to compact subsets of Q.

Thus, if we set $K = \{\overline{\mu} \in \mathbf{CConv}(Q)^* \mid \mu \in Q^*\}$ then all the preconditions on the set K in the previous section are satisfied: By the above argument the intersection of K with the polar of each neighborhood is $w(\mathbf{CConv}(Q)^*, \mathbf{CConv}(Q))$-compact, and as all elements of Q are bounded, K is strictly separating for $\mathbf{CConv}(Q)$; more precisely: For all $v \in V$, the corresponding neighborhood $\overline{v} \in \overline{V}$ and $A,B \in \mathbf{CConv}(Q)$

$$A \not\leq B+\overline{v} \text{ implies } \overline{\mu}(A) > \overline{\mu}(B)+1 \text{ for some } \overline{\mu} \in K \cap \overline{v}^\circ_{\mathbf{CConv}(Q)}\,.$$

We are now going to apply Theorem 2.5 in order to derive a Korovkin type theorem for set-valued functions generalizing the results in [59] and [28] which only consider compact domains and finite dimensional vector spaces. To prepare our main result we need to introduce the following notations:

3.1 Half spaces. A subset H of a vector space Q is said to be a *half space* if both H and its complement $Q\backslash H$ are convex.

It easy to check (using the classical Hahn-Banach Separation Theorem) that the half spaces H of Q which are both decreasing and closed with respect to the symmetric topology all are of the type

$$H = \{a \in Q \mid \mu(a) \leq \alpha\} \text{ for some } \mu \in Q^* \text{ and } \alpha \in \mathbb{R}.$$

3.2 Absorbing and saturating families. A subfamily $\boldsymbol{E}$ of $\boldsymbol{CConv(Q)}$ is said to be *absorbing* if for every $a \in Q$ there is some E in the subcone of $\boldsymbol{CConv(Q)}$ generated by $\boldsymbol{E}$ such that $0, a \in E$.

A subfamily $\boldsymbol{C}$ of $\boldsymbol{CConv(Q)}$ is said to be *saturating* if for every closed and decreasing half space H in Q, every element $a \in H$ and $b \in \overset{\circ}{H}$ (the topological interior of H with respect to the symmetric topology) there is some C in the subcone of $\boldsymbol{CConv(Q)}$ generated by $\boldsymbol{C}$ such that $a, b \in C \subset H$.

As all elements of $\boldsymbol{CConv(Q)}$ are bounded we may deal with their negatives as well and consider the locally convex cone $\boldsymbol{CConv(Q)}$-$\boldsymbol{CConv(Q)}$ of all formal differences of elements of $\boldsymbol{CConv(Q)}$ (c.f Ch. I.4.8). It has the same dual as $\boldsymbol{CConv(Q)}$, and all u-continuous operators on $\boldsymbol{CConv(Q)}$ may be extended to $\boldsymbol{CConv(Q)}$-$\boldsymbol{CConv(Q)}$ in a canonical way (c.f. Ch. II.1.5). We have to be careful, however, not to confuse the formal negative of the element $A \in \boldsymbol{CConv(Q)}$ with the subset $\{-a \mid a \in A\}$. The element $-A$ is just the additive inverse of A in $\boldsymbol{CConv(Q)}$-$\boldsymbol{CConv(Q)}$ and may, except for singleton sets, not be identified with a subset of Q.

3.3 Theorem. *Suppose that the separated locally convex cone (Q,V) is a vector space. Let X be a locally compact Hausdorff space, Ψ a family of weights on X, M a ${}^{+}$Korovkin system for $C\Psi(X)$ and $p \in C\Psi(X)$ a strictly positive function. Let $\boldsymbol{E}$ be an absorbing, $\boldsymbol{C}$ a saturating subfamily of $\boldsymbol{CConv(Q)}$.*
Let $\boldsymbol{C_0}$ be a subcone of $\boldsymbol{C\Psi(X,CConv(Q)\text{-}CConv(Q))}$ such that all the functions

$$x \to g(x)E \;:\; X \to Q, \quad g \in M,\ E \in \boldsymbol{E}$$

and

$$x \to p(x)C \;:\; X \to Q, \quad C \in \boldsymbol{C}$$

are contained in $\boldsymbol{C_0}$.
Then every set-valued continuous function $f \in \boldsymbol{C\Psi(X,CConv(Q)\text{-}CConv(Q))}$ is $\boldsymbol{C_0}$-superharmonic in the elements $\overline{\mu}_x$, for all $\mu \in Q^*$ and all $x \in X$ such that $\psi(x) \neq 0$ for some $\psi \in \Psi$.

Proof. We only have to verify that the above conditions yield the criteria of Theorem 2.5. Note that a set $A \in \boldsymbol{CConv(Q)}$ is positive if and only if $0 \in A$. Clearly, we may assume that the collections $\boldsymbol{E}$ and $\boldsymbol{C}$ are already subcones of $\boldsymbol{CConv(Q)}$. Thus, (i) in 2.5 follows immediately from the first of the above sets of functions included in $\boldsymbol{C_0}$. Now let $\mu, \tau \in Q^*$ such that $\overline{\tau}$ is not a multiple of $\overline{\mu}$ in $\boldsymbol{CConv(Q)}^* = (\boldsymbol{CConv(Q)\text{-}CConv(Q)})^*$. Then τ is not a multiple of μ in Q^*. A simple argument shows that the topological interior of the closed and decreasing half space $H = \{a \in Q \mid \mu(a) \leq 0\}$ contains an element b such that $\tau(b) > 0$. Thus, by our assumption (ii) there is a set $C \in \boldsymbol{C}$ such that $0, b \in C \subset H$; i.e. $C \geq 0$, and we have $\overline{\tau}(C) \geq \tau(b) > \overline{\mu}(C) = 0$. Thus, condition (ii) of Theorem 2.5 is satisfied as well. So all left to show is that for every $x \in X$ and $0 \neq \mu \in Q^*$ there are functions $g, h \in \boldsymbol{C_0}$ such that $\overline{\mu}(g(x)) > 0$ and $\overline{\mu}(h(x)) < 0$. But this follows immediately using the assumption of our theorem for the half spaces $G = \{a \in Q \mid \mu(a) \geq 1\}$ and $H = \{a \in Q \mid \mu(a) \leq -1\}$, elements $G \supset C_G \in \mathcal{C}$ and $H \supset C_H \in \mathcal{C}$ and the respective functions $x \to p(x)C_G$ and $x \to p(x)C_H$ in $\mathcal{C}_0$.

3.4 Examples. (a) If the subset $\mathcal{C}$ of $CConv(Q)$ contains all compact segments of Q; i.e. the sets $\{\lambda a+(1-\lambda)b \mid 0 \leq \lambda \leq 1\}$ for $a,b \in Q$, then it is saturating. Every saturating subfamily of $CConv(Q)$ is obviously absorbing.

(b) For $Q = \mathbb{R}^n$ with the Euclidean unit ball B choose the absorbing family $\mathcal{E} = \{B\}$. Furthermore, if $\mathcal{C} \subset CConv(Q)$ contains B as well as the singleton sets $\{e_k\}$ and $\{-e_k\}$ running through a basis $\{e_1,...,e_n\}$ of $\mathbb{R}^n$, then it is easily seen to be saturating. But notice that B cannot be replaced by the cubical unit ball $B' = [-1,+1]^n$.

Combining Theorem 3.3 with the Convergence Theorem IV.1.11 yields a substantial generalization of the results about set-valued functions in [59] and [28]. We shall give two versions. The first one deals only with operators converging to the identity. For the following convergence results we shall restrict ourselves to the cone $C\Psi(X,CConv(Q))$ and avoid the formal negatives of sets. This requires, however, a closer look on convergence statements in case that our $^+$Korovkin system M contains non-positive functions: For a function $g \in C\Psi(X)$ we denote as usual by $g^+ = g \vee 0$ and $g^- = (-g) \vee 0$ its positive and its negative parts. Both g^+ and g^- are in $C\Psi(X)$, and $g = g^+ - g^-$. We have to deal with convergence statements of the type $T_\alpha(g \cdot E) \uparrow g \cdot E$, where $g \in M$, $E \in CConv(Q)$ and $(T_\alpha)_{\alpha \in A}$ is a u-equicontinuous net of linear operators on $C\Psi(X,CConv(Q))$. Recall that $g \cdot E$ stands for the function $x \to g(x)E : X \to Q$. Convergence in the upper, lower and symmetric topologies was denoted by the symbols $\uparrow$, $\downarrow$ and $\to$, respectively. As we want to formulate our convergence conditions in terms of positive functions and elements of $CConv(Q)$ only, we shall require that

$$T_\alpha(g^+ \cdot E) \uparrow g^+ \cdot E \quad \text{and} \quad T_\alpha(g^- \cdot E) \downarrow g^- \cdot E \,.$$

The latter is equivalent to $T_\alpha((-g^-) \cdot E) \uparrow (-g^-) \cdot E$, whence both conditions imply

$$T_\alpha(g \cdot E) \uparrow g \cdot E \,,$$

indeed. The following is obvious if we combine Theorem 3.3 with the Convergence Theorem IV.1.11:

3.5 Corollary. *Suppose that the separated locally convex cone $(Q,\mathcal{V})$ is a vector space. Let X be a locally compact Hausdorff space, Ψ a family of weights on X, M a $^+$Korovkin system for $C\Psi(X)$ and $p \in C\Psi(X)$ a strictly positive function. Let $\mathcal{E}$ be an absorbing, $\mathcal{C}$ a saturating subfamily of $CConv(Q)$.*

Let $(T_\alpha)_{\alpha \in A}$ be a u-equicontinuous net of linear operators on $C\Psi(X,CConv(Q))$. If

$$T_\alpha(g^+ \cdot E) \uparrow g^+ \cdot E \quad \text{and} \quad T_\alpha(g^- \cdot E) \downarrow g^- \cdot E \quad \text{for all} \quad g \in M \text{ and } E \in \mathcal{E}$$

and
$$T_\alpha(p \cdot C) \uparrow p \cdot C \quad \text{for all} \quad C \in \mathcal{C},$$

then
$$T_\alpha(f) \to f \quad \text{for all} \quad f \in C\Psi(X,CConv(Q));$$

i.e. if for all $g \in M$, $E \in \mathbf{E}$, $C \in \mathbf{C}$, $v \in V$ *and* $\psi \in \Psi$ *there is* $\alpha_0 \in A$ *such that*

$$\psi(x)T_\alpha(g^+ \cdot E)(x) \leq \psi(x)g^+(x)E+\bar{v}, \quad \psi(x)g^-(x)E \leq \psi(x)T_\alpha(g^- E)(x)+\bar{v}$$

and

$$T_\alpha(p\cdot C)(x) \leq p(x)C+\bar{v}$$

for all $x \in X$ *and* $\alpha \geq \alpha_0$, *then for every* $f \in \mathbf{C}\Psi(X,\mathbf{CConv}(Q))$, $v \in V$ *and* $\psi \in \Psi$ *there is* $\alpha_0 \in A$ *such that*

$$\psi(x)T_\alpha(f)(x) \leq \psi(x)f(x)+\bar{v} \quad \textit{and} \quad \psi(x)f(x) \leq \psi(x)T_\alpha(f)(x)+\bar{v}$$

for all $x \in X$ *and* $\alpha \geq \alpha_0$.

3.6 Examples. For $Q = \mathbf{R}^n$ with the Euclidean unit ball B, let $\mathbf{E} = \{B\}$ and $\mathbf{C} = \{B, \{\pm e_1\},\ldots,\{\pm e_n\}\}$ as in 3.4(b). In this case the condition on the operators T_α reads as follows:

$$T_\alpha(g^+\cdot B) \uparrow g^+\cdot B \quad \text{and} \quad T_\alpha(g^-\cdot B) \downarrow g^-\cdot B \quad \text{for all } g \in M,$$
$$T_\alpha(p\cdot B) \uparrow p\cdot B \quad \text{and} \quad T_\alpha(\{e_k\}) \to \{e_k\} \quad \text{for all } k = 1,\ldots,n.$$

If this is satisfied for a net (or sequence) of positive linear operators on $\mathbf{C}(X,\mathbf{CConv}(\mathbf{R}^n))$, then $T_\alpha(f) \to f$ for every $f \in \mathbf{C}(X,\mathbf{CConv}(\mathbf{R}^n))$.

For $X = [0,1]$ and $\Psi = \{1\}$, to give an example, we may consider the $^+$Korovkin system generated by the three functions (c.f. Example 2.3(b)) $f_0(x) = 1$, $f_1(x) = -x$, and $f_2(x) = x^2$.

For $X = \mathbf{R}$, but now Ψ consisting of the characteristic functions of the compact subsets of $\mathbf{R}$, i.e $\mathbf{C}\Psi(X) = \mathbf{C}(\mathbf{R})$ with the topology of compact convergence, choose the same functions as above (Example 2.3(b)).

For $X = \mathbf{R}$ and $\Psi = \{1\}$ we may use the $^+$Korovkin system as in Example 2.3(c) i.e. the functions $f_0(x) = e^{-x^2}$, $f_1(x) = -xe^{-x^2}$, and $f_2(x) = x^2e^{-x^2}$.

Now we may turn to the case of more general operators S_α: With the above notations for (Q,V), $\mathbf{C}\Psi(X,\mathbf{CConv}(Q))$ and $\mathbf{C_0}$ let (P,W) be another locally convex cone. We consider equicontinuous nets $(T_\alpha)_{\alpha \in A}$ of monotone linear operators from $\mathbf{C}\Psi(X,\mathbf{CConv}(Q))$ into $\mathbf{C}\Psi(X,\mathbf{CConv}(P))$. According to the General Convergence Theorem IV. 1.13 we are looking for appropriate operators $(S_\alpha)_{\alpha \in A}$ in order to study convergence: By Example 1.6(e) from Ch. II any u-continuous linear operator $S : Q \to P$ may be extended in a canonical way into a u-continuous operator $\bar{S} : \mathbf{CConv}(Q) \to \mathbf{CConv}(P)$. (Recall: $\bar{S}(A) = \overline{S(A)}$, i.e. the closure of $S(A) = \{S(a) \mid a \in A\}$ with respect to the lower topology, for $A \in \mathbf{CConv}(Q)$.) Those extensions were even seen to be up-directional (Remark II.6.2) with respect to the $\vee$-semilattice structure of $\mathbf{CConv}(Q)$ and $\mathbf{CConv}(P)$. Finally, we associate with S an operator $\tilde{S}$ from $\mathbf{C}\Psi(X,\mathbf{CConv}(Q))$ into $\mathbf{C}\Psi(X,\mathbf{CConv}(P))$ defined by

$$(\tilde{S}(f))(x) = \bar{S}(f(x)) = \overline{S(f(x))} \quad \text{for all } f \in \mathbf{C}\Psi(X,\mathbf{CConv}(Q)).$$

It is easily checked that for elements $\tau \in P^*$ and $x \in X$ this yields $\tilde{S}(\bar{\tau}_x) = \overline{(S^*(\tau))}_x$. Thus, for equicontinuous nets $(S_\alpha)_{\alpha \in A}$ of linear operators from Q into P and the corresponding net $(\tilde{S}_\alpha)_{\alpha \in A}$ of operators from $\mathbf{C}\Psi(X,\mathbf{CConv}(Q))$ into $\mathbf{C}\Psi(X,\mathbf{CConv}(P))$ the condition from the General Convergence Theorem IV.1.13 (the subset $\bigcap_{\alpha\in A}\tilde{S}_\alpha^{*-1}(K)$ of P^* is strictly separating for $C\Psi(X,CConv(P))$, for $K=\{\bar{\mu}_x \mid \mu\in Q^*,\ x\in X$ such that $\psi(x)\neq 0$ for some $\psi\in\Psi\}$, is obviously satisfied. Summarizing, we reformulate Corollary 3.5:

3.7 Corollary. *Suppose that* (Q,V) *and* (P,W) *are separated locally convex cones and that* Q *is even a vector space. Let* X *be a locally compact Hausdorff space,* Ψ *a family of weights on* X, M *a* $^+$*Korovkin system for* $C\Psi(X)$ *and* $p\in C\Psi(X)$ *a strictly positive function. Let* $\mathbf{E}$ *be an absorbing,* $\mathbf{C}$ *a saturating subfamily of* $CConv(Q)$.
Let $(S_\alpha)_{\alpha\in A}$ *be a u-equicontinuous net of linear operators from* Q *into* P *and* $(T_\alpha)_{\alpha\in A}$ *a u-equicontinuous net of linear operators from* $C\Psi(X,CConv(Q))$ *into* $C\Psi(X,CConv(P))$. *If for every* $g\in M$, $E\in\mathbf{E}$, $C\in\mathbf{C}$, $w\in W$ *and* $\psi\in\Psi$ *there is* $\alpha_0\in A$ *such that*

$$\psi(x)T_\alpha(g^+\cdot E)(x) \le \psi(x)g^+(x)\overline{S_\alpha(E)}+\bar{w},$$

$$\psi(x)g^-(x)\overline{S_\alpha(E)} \le \psi(x)T_\alpha(g^-\cdot E)(x)+\bar{w}$$

and

$$T_\alpha(p\cdot C)(x) \le p(x)\overline{S_\alpha(C)}+\bar{v}$$

for all $x\in X$ *and* $\alpha\ge\alpha_0$. *Then this implies already convergence with respect to the uniform symmetric topology for all functions in* $C\Psi(X,CConv(Q))$; *more precisely: For every* $f\in C\Psi(X,CConv(Q))$, $w\in W$ *and* $\psi\in\Psi$ *there is* $\alpha_0\in A$ *such that*

$$\psi(x)T_\alpha(f)(x) \le \psi(x)\overline{S_\alpha(f(x))}+\bar{w} \quad\text{and}\quad \psi(x)\overline{S_\alpha(f(x))} \le \psi(x)T_\alpha(f))(x)+\bar{w}$$

for all $x\in X$ *and* $\alpha\ge\alpha_0$.

We shall proceed with a Stone-Weierstraß type theorem for set-valued functions. The following prepares our result:

3.8 Selection Lemma. *Suppose that the separated locally convex cone* (Q,V) *is a vector space. Let* X *be a locally compact Hausdorff space,* Ψ *a family of weights on* X *and* $f\in C\Psi(X,CConv(Q))$. *For every* $x\in X$, $a\in f(x)$, $v\in V$ *and* $\psi\in\Psi$ *there is a single-valued function* $h\in C\Psi(X,Q)$ *such that*

$$h(x)=a \quad\text{and}\quad \psi(y)\overline{h(y)} \le \psi(y)f(y)+\bar{v} \quad\text{for all } y\in X.$$

Moreover, all the values of h *are contained in some finite dimensional subspace of* Q.

Proof. Let f,x,a and v and ψ be as in the lemma. By our Definition 1.2 of a Nachbin cone there is a compact subset Y of X containig the element a such that

$$\psi(x)f(x)\le\bar{v} \quad\text{and}\quad 0\le\psi(x)f(x)+\bar{v} \text{ for all } x\in X\setminus Y.$$

Now for every $y\in Y$ select an element $a_y\in f(y)$. Set $a_x=a$ in particular. As the function f is continuous, for every $y\in Y$ we find an open neighborhood U_y of y such that both

$$\overline{a_y}\le f(z)+(1/(\psi(y)+1))\bar{v} \quad\text{and}\quad \psi(z)\le\psi(y)+1 \quad\text{for all } z\in U_y.$$

(Recall that the bar over a_y refers to the closure with respect to the lower topology (c.f. Ch. I.3.3) which renders an element of $CConv(Q)$).) This yields

$$\psi(z)\overline{a_y} \le \psi(z)f(z)+(\psi(z)/(\psi(y)+1))\bar{v}\le\psi(z)f(z)+\bar{v} \quad\text{for all } z\in U_y.$$

The sets U_y together with $U_0 = X\setminus Y$ cover the one-point compactification $\overline{X} = X\cup\{\infty\}$ of X, whence admit a finite subcovering $\{U_0, U_{y_1},...,U_{y_n}\}$. In addition, we may arrange that $y_1 = x$ and that x is contained in no other set of this covering but U_{y_1}. Let $\{\phi_0,\phi_1,...,\phi_n\}$ be a corresponding partition of the unit, i.e. a collection of positive continuous real-valued functions such that $\phi_0+\phi_1+...+\phi_n = 1$ on $\overline{X}$, and every ϕ_i vanishes outside the set U_{y_i}. Finally, set

$$h = \phi_1 a_{y_1} + \phi_2 a_{y_2} + ... + \phi_n a_{y_n} \in C\Psi(X,Q).$$

Then we have $\overline{h(y)} \le f(y)+\bar{v}$ for all $y \in X$, and as $\phi_1(x) = 1$ and $\phi_2(x) = ... = \phi_n(x) = 0$, this shows $h(x) = a_{y_1} = a$ as well. Moreover, the values of h are all contained in the subspace of Q spanned by the elements $a_{y_1},..., a_{y_n}$.

3.9 Remarks. (a) Note that in the proof of Lemma 3.8 we only use that the set-valued function f is continuous with respect to the lower topology on $CConv(Q)$. Thus, a similar statement holds for lower semicontinuous $CConv(Q)$-valued functions.

(b) A single-valued function $h \in C(X,Q)$ such that $h(y) \in f(y)$ for all $y \in X$ and a set-valued function $f \in C(X,CConv(Q))$, is usually called a *continuous selection* of f. Continuous selections have been dealt with in various places, in particular detail by E. Michael in [32] and[33]. They are known to exist, though not in general, but in some important special cases: for example if Q is a complete metric space.

3.10 Theorem. *Suppose that the separated locally convex cone (Q,V) is a vector space. Let X be a locally compact Hausdorff space and Ψ a family of weights on X. For a finite set of single-valued functions $f_1,...,f_n \in C\Psi(X,Q)$ which are all of finite rank, i.e. their values are contained in some finite dimensional subspace of Q, consider the set-valued function*

$$x \to \overline{conv}\{f_1(x),...,f_n(x)\} \ : \ X \to CConv(Q).$$

(closure is meant with respect to the lower topology.) Then the subcone C_0 of all those functions is dense in $C\Psi(X,CConv(Q))$ with respect to the symmetric topology.

Proof. Clearly, C_0 is sup-stable. Thus, by Proposition III. 2.3 it is a uniformly up-directed subcone of $C\Psi(X,CConv(Q))$. Using Proposition 1.9 of this chapter, all left to show is that every function in $C\Psi(X,CConv(Q))$ is C_0-subharmonic in all elements $\overline{\mu}_x \in C\Psi(X,CConv(Q))^*$, $\mu \in Q^*$, $x \in X$. We shall use the "if"-part of the Sup-Inf-Theorem III.1.3 and the preceding lemma in order to verify this:
For $f \in C\Psi(X,CConv(Q))$ and $\overline{\mu}_x \in C\Psi(X,CConv(Q))^*$ by the definition of $\overline{\mu}_x$ and by the compactness of $f(x)$ there is some $a \in f(x)$ such that $\mu(a) = \overline{\mu}_x(f)$. Now, corresponding to Theorem III.1.3, let $\bar{v}_\psi \in \overline{V}_\Psi$ and choose the single-valued function $h \in C\Psi(X,Q)$ as in Lemma 3.8. Clearly, h is contained in C_0 (we identify its single values with their closure), and we have

$$h \le f+\bar{v}_\psi \quad \text{and} \quad \overline{\mu}_x(h) = \overline{\mu}_x(f),$$

thus guaranteeing the condition in III.1.3.

We shall conclude this section with an approximation theorem for functions with values in the locally convex cone $\overline{BConv(Q)}$ of all non-empty bounded decreasing and lower closed convex subsets of Q, thus recovering and generalizing results by Prolla in [44]. As we mentioned in the introduction to this section, we will not be able to identify a suitable subset K of the dual of this larger cone. Thus, we have to consider the whole dual $\overline{BConv(Q)}^*$ instead and use our less powerful Theorem 2.4 in order to obtain the desired approximation result. For the first result on subharmonicity we shall use the locally convex cone $\overline{BConv(Q)}\text{-}\overline{BConv(Q)}$.

3.11 Theorem. *Suppose that the separated locally convex cone* (Q,V) *is a vector space. Let* X *be a locally compact Hausdorff space,* Ψ *a family of weights on* X, M *a* $^+$*Korovkin system for* $C\Psi(X)$ *and* $p \in C\Psi(X)$ *a strictly positive function. Let* $\mathbf{E}$ *be a subset of* $\overline{BConv(Q)}$ *such that for all* $A \in \overline{BConv(Q)}$ *and* $\bar{v} \in \overline{V}$ *there is some* $E \in \mathbf{E}$ *and* $\rho > 0$ *such that* $A \subset \rho E + \bar{v}$. *Let* C_0 *be a subcone of* $C\Psi(X, \overline{BConv(Q)}\text{-}\overline{BConv(Q)})$ *such that all the functions*

$$x \to g(x)E : X \to Q, \quad g \in M, \ E \in \mathbf{E}$$

and

$$x \to p(x)C : X \to Q, \quad C \in \overline{BConv(Q)}\text{-}\overline{BConv(Q)}$$

are contained in C_0.
Then every function $f \in C\Psi(X, \overline{BConv(Q)}\text{-}\overline{BConv(Q)})$ is C_0-*superharmonic on* $\overline{BConv(Q)}^*{}_{\Psi X}$.

For the corresponding convergence result we shall again restrict ourselves to the cone $C\Psi(X, \overline{BConv(Q)})$ and refer to the remarks preceding Corollary 3.4:

3.12 Corollary. *Suppose that the separated locally convex cone* (Q,V) *is a vector space. Let* X *be a locally compact Hausdorff space,* Ψ *a family of weights on* X, M *a* $^+$*Korovkin system for* $C\Psi(X)$ *and* $p \in C\Psi(X)$ *a strictly positive function. Let the subset* $\mathbf{E} \subset \overline{BConv(Q)}$ *be as in Theorem* 3.11.
Let $(T_\alpha)_{\alpha \in A}$ *be a u-equicontinuous net of linear operators on* $C\Psi(X, \overline{BConv(Q)})$. *If*

$$T_\alpha(g^+ \cdot E) \uparrow g^+ \cdot E \quad \text{and} \quad T_\alpha(g^- \cdot E) \downarrow g^- \cdot E \quad \text{for all} \quad g \in M \ \text{and} \ E \in \mathbf{E}$$

and

$$T_\alpha(p \cdot C) \to p \cdot C \quad \text{for all} \quad C \in \overline{BConv(Q)},$$

then

$$T_\alpha(f) \to f \quad \text{for all} \quad f \in C\Psi(X, \overline{BConv(Q)});$$

i.e. if for all $g \in M$, $E \in \mathbf{E}$, $C \in \overline{BConv(Q)}$, $v \in V$ *and* $\psi \in \Psi$ *there is* $\alpha_0 \in A$ *such that*

$$\psi(x)T_\alpha(g^+ \cdot E)(x) \le \psi(x)g^+(x)E + \bar{v}, \quad \psi(x)g^-(x)E \le \psi(x)T_\alpha(g^- \cdot E)(x) + \bar{v}$$

and

$$T_\alpha(p \cdot C)(x) \le p(x)C + \bar{v} \ \text{and} \ p(x)C \le T_\alpha(p \cdot C)(x) + \bar{v}$$

for all $x \in X$ *and* $\alpha \ge \alpha_0$, *then for every* $f \in C\Psi(X, \overline{BConv(Q)})$, $v \in V$ *and* $\psi \in \Psi$ *there is* $\alpha_0 \in A$ *such that*

$$\psi(x)T_\alpha(f)(x) \le \psi(x)f(x) + \bar{v} \ \text{and} \ \psi(x)f(x) \le \psi(x)T_\alpha(f)(x) + \bar{v}$$

for all $x \in X$ *and* $\alpha \ge \alpha_0$.

Note that the condition in Corollary 3.12 is considerably stronger than in 3.5, in particular as it requires all (replacements for the) constant functions $p \cdot C$ to be contained in the "test" set for the convergence of the operators.

We conclude this chapter by applying our last corollary to normed vector spaces, a case of particular interest:

3.13 Corollary. *Let* $(E, \|\ \|)$ *be a normed space with unit ball* B. *Let* X *be a locally compact Hausdorff space,* Ψ *a family of weights on* X, M *a* $^{+}$*Korovkin system for* $C\Psi(X)$ *and* $p \in C\Psi(X)$ *a strictly positive function. Let* $(T_\alpha)_{\alpha \in A}$ *be a u-equicontinuous net of linear operators on* $C\Psi(X, \overline{BConv(E)})$. *If*

$$T_\alpha(g^+ \cdot B) \uparrow g^+ \cdot B \quad \text{and} \quad T_\alpha(g^- \cdot B) \downarrow g^- \cdot B \quad \text{for all} \quad g \in M$$

and

$$T_\alpha(p \cdot C) \to p \cdot C \quad \text{for all} \quad C \in \overline{BConv(E)}),$$

then

$$T_\alpha(f) \to f \quad \text{for all} \quad f \in C\Psi(X, \overline{BConv(E)}).$$

Chapter VI: Quantitative Estimates

In Chapter V we proved several Korovkin type approximation theorems of qualitative nature. The problem was to exhibit small sets M of "test functions" in the function spaces under consideration such that, for every equicontinuous sequence $(T_n)_{n \in N}$ of monotone linear operators on the function space, $T_n(f)$ converges to f provided that this holds for the test functions $g \in M$. We now turn our attention to the quantitative aspects of the convergence. In Section 4 of this chapter we present several results on the order of convergence: Knowing the order of convergence of $T_n(g)$ towards g for the test functions $g \in M$, we derive results on the order of convergence of $T_n(f)$ towards f for arbitrary functions f depending on their modulus of continuity and the "smoothness" of their values relative to those of the test functions. For this we have to extend the classical notions of approximation theory like order of convergence, modulus of continuity to our general setting of locally convex cones. This is done in the first three sections. Our basic result is Theorem 4.1. Because of its generality, its formulation is rather lengthy and complicated. So is its proof. The Corollaries 4.3, 4.5, 4.6, 4.9 and 4.10 for weighted spaces of real-valued, complex-valued, various types of set-valued functions and stochastic processes justify our efforts. The first of these generalizes the well-known quantitative Korovkin type theorems of de Vore [18] for real-valued functions in one real variable that had been generalized to several real variables by Censor [15] and to functions on compact spaces by Nishishiraho [35], [36], [37].

1. Sequence cones.

The concept of sequence cones which we shall introduce in the following will provide our main tool for obtaining estimates on the order of convergence for Korovkin type approximation in locally convex cones. As usual, let (Q,V) be a locally convex cone. For sequences $(a_n)_{n \in N}$ and $(b_n)_{n \in N}$ of elements of Q and a neighborhood $v \in V$ we define a corresponding neighborhood which, by abuse of language, we shall denote by the same symbol:

$$(a_n) \leq (b_n)+v \text{ if and only if } a_n \leq b_n+v \text{ for all } n \in N.$$

Thus, V becomes an abstract neighborhood system for the cone of sequences in Q. (The algebraic operations for sequences, of course, are defined componentwise.) Finally, by Q^N let us denote the cone of those sequences which are uniformly bounded below, i.e. for all $(a_n) \in Q^N$ and all $v \in V$ there is some $\rho \geq 0$ such that $0 \leq a_n + \rho v$ for all $n \in N$.

Thus, (Q^N, V) becomes a locally convex cone to which we shall refer in the sequel as the *sequence cone over* (Q,V).

Although N is locally compact in its usual (discrete) topology and sequences may be considered as functions defined on N, the sequence cone (Q^N, V) admits no representation as a Nachbin cone in the sense of Chapter V, as we do not impose the condition for sequences to vanish at infinity. Unboundedness at infinity will in fact turn out to be essential for our applications. They justify our choice of not restricting our concept of a locally convex cone to the case where all elements are bounded, case in which they would be embeddable in locally convex vector spaces. Also for sequence cones we have to deal with the general notation of super- and subharmonicity (IV.1.8) on subsets of the dual cone rather than on single elements, as the preconditions of Proposition IV.1.10 will not be available in this situation, and we shall need the general version of our Convergence Theorem IV.1.13. For this, we shall proceed to identify suitable subsets K of the dual Q^{N*}.

For a u-continuous linear functional $\mu \in Q^*$ and a fixed $i \in N$ we shall denote by μ^i the linear functional on Q^N defined by

$$\mu^i((a_n)_{n \in N}) = \mu(a_i) \quad \text{for all } (a_n)_{n \in N} \in Q^N.$$

Clearly, we have $\mu^i \in (v)^\circ_{Q^N}$ whenever $\mu \in v^\circ_Q$. If K is a subset of Q^* we shall denote by $K_i = \{\mu^i \mid \mu \in K\}$ and by $K_N = \bigcup_{i \in N} K_i$.

The following is a first observation:

1.1 Lemma. *If the subset* K *of* Q^* *is strictly separating for* Q, *then the subset* K_N *of* Q^{N*} *is strictly separating for* Q^N.

Proof. By the assumption on K, for every neighborhood $v \in V$, there is a u-equicontinuous subset K_0 of K and some $\rho > 0$ such that for all $a, b \in Q$

$$a \not\leq b + v \quad \text{implies} \quad \mu(a) > \mu(b) + \rho \quad \text{for some } \mu \in K_0.$$

The subset $(K_0)_N = \{\mu^i \mid \mu \in K_0,\ i \in N\}$ of Q^{N*} is clearly u-equicontinuous as well. Now let (a_n) and (b_n) be elements of Q^N such that

$$(a_n) \not\leq (b_n) + v, \quad \text{i.e.} \quad a_i \not\leq b_i + v \quad \text{for some } i \in N.$$

But then there is some $\mu \in K_0$ such that $\mu(a_i) > \mu(b_i) + \rho$, whence $\mu^i((a_n)) > \mu^i((b_n))$ for the element $\mu^i \in (K_0)_N$.

The topological properties of the subset K of Q^*, however, as required in Proposition IV.1.10 will not transfer to K_N in general. So we have to deal with the general definition of super- and subharmonicity via nets (resp. sequences) as in IV.1.8. The main result of this section is a criterion similar to the "if"-part of the Sup-Inf-Theorem III.1.3 which will allow us to handle this problem:

1.2 Proposition. *Let (Q,V) be a locally convex cone and K a subset of Q^* such that the intersection of K with the polar of each neighborhood is closed. Let $(Q^N)_0$ be a subcone of the sequence cone Q^N. For the element $(a_n)\in Q^N$ suppose that $\mu(a_i)$ is finite for all $\mu\in K$ and for all $i\in N$. Suppose that the following holds:*
For every u-equicontinuous sequence (μ_n) in K, every $v\in V$ and $\varepsilon>0$ there is an element $(b_n)\in(Q^N)_0$ such that

$$a_n\le b_n+v \quad \text{and} \quad \mu_n(b_n)\le\mu_n(a_n)+\varepsilon \quad \text{for all} \quad n\in N.$$

Then (a_n) is sequentially $(Q^N)_0$-superharmonic on K_N.
An analogous statement holds for $(Q^N)_0$-subharmonicity.

Proof. We have to check carefully the condition of Definition IV.1.8: Let us consider u-equicontinuous sequences (τ_m) in K_N and (σ_m) in Q^{N*} such that

$$(\sigma_m((b_n)))\precsim(\tau_m((b_n))) \quad \text{for all } (b_n)\in(Q^N)_0;$$

i.e. for every $(b_n)\in(Q^N)_0$ and $\varepsilon\ge0$ there is some $m_0\in N$ such that

$$\sigma_m((b_n))\le\tau_m((b_n))+\varepsilon \quad \text{for all } m\ge m_0.$$

By the way of contradiction, let us suppose that $(\sigma_m((a_n)))\precsim(\tau_m((a_n)))$ does not hold, i.e. there is some $\delta>0$ such that (after selecting an appropriate subsequence) we have $\sigma_m((a_n))\ge\tau_m((a_n))+\delta$ for all $m\in N$.
Now let us denote by $\phi:N\to N$ the mapping defined by $\tau_m\in K_{\phi(m)}$.

Firstly, suppose that there is some $i\in N$ such that $\phi^{-1}(i)$ contains infinitely many elements. Thus, by selecting an appropriate subnet $(\tau_{m_\alpha})_{\alpha\in A}$, we may assume that there is a convergent net $(\mu_\alpha)_{\alpha\in A}$ in $K\subset Q^*$ such that

$$\tau_{m_\alpha}=(\mu_\alpha)^i\in K_i \quad \text{for all } \alpha\in A.$$

Furthermore, our assumption on K implies that $\lim_\alpha\mu_\alpha=\mu\in K$. Now set $\varepsilon=\delta/8$ and let $v\in V$ be such that both sequences (τ_n) and (σ_n) are contained in $\varepsilon(v)^\circ_{Q^N}$. Now choose a sequence $(b_n)\in(Q^N)_0$ as in the proposition corresponding to the stationary sequence (μ) in K. This shows that for some $\alpha_0\in N$ and all $\alpha\ge\alpha_0$ we have

$$\sigma_{m_\alpha}((a_n))\le\sigma_{m_\alpha}((b_n))+\varepsilon\le\tau_{m_\alpha}((b_n))+2\varepsilon=\mu_{m_\alpha}(b_i)+2\varepsilon\le\mu(b_i)+3\varepsilon\le\mu(a_i)+4\varepsilon.$$

On the other hand we assumed $\sigma_{m_\alpha}((a_n))\ge\tau_{m_\alpha}((a_n))+\delta=\mu_{m_\alpha}(a_i)+\delta$ for all $\alpha\ge\alpha_0$, whence $\sigma_{m_\alpha}((a_n))>\mu(a_i)+\delta/2=\mu(a_i)+4\varepsilon$ for sufficiently large $\alpha\in N$, a clear contradiction.

Now secondly, suppose that for all $i\in N$ the set $\phi^{-1}(i)$ contains only finitely many elements. By selecting an appropriate subsequence, we may assume that every $\phi^{-1}(i)$ contains at most one element. Furthermore, we can find a sequence (μ_n) in K such that (τ_m) is a subsequence of $((\mu_n)^n)$ in K_N. Again, set $\varepsilon=\delta/4$ and let $v\in V$ be such that both sequences (τ_n) and (σ_n) are contained in $\varepsilon(v)^\circ_{Q^N}$ and choose the sequence $(b_n)\in(Q^N)_0$ as in the proposition corresponding to the sequence (μ_n) in K. This shows for all $m\ge m_0$ and the corresponding indices n for the sequence (μ_n) that

$$\sigma_m((a_n)) \le \sigma_m((b_n))+\varepsilon \le \tau_m((b_n))+2\varepsilon = \mu_n(b_n)+2\varepsilon \le \mu_n(a_n)+3\varepsilon,$$

contradicting $\sigma_m((a_n)) \ge \tau_m((a_n))+\delta = \mu_n(a_n)+\delta = \mu_n(a_n)+4\varepsilon$ for all $m \ge m_0$. This completes our proof.

2. Order of convergence for Korovkin type approximation.

We are now going to study the order of convergence for Korovkin type approximation processes in locally convex cones. We consider only sequences of operators and use the techniques developed in the previous section in order to obtain our results.

2.1 Order of convergence. Equiconvergence. Let (Q,V) and (P,W) be two locally convex cones and W' a basis for the abstract neighborhood system W. Let $(T_n, S_n)_{n \in N}$ be a pair of sequences of u-equicontinuous linear operators from Q into P. For a subset A of Q and a sequence (α_n) of strictly positive numbers converging to zero we shall say that $(T_n,S_n)_{n \in N}$ *is equiconvergent on* A *of order* $o(\alpha_n)$ if and only if for every $w \in W'$ and every $\varepsilon > 0$ there is some $n_0 \in N$ such that

$$T_n(a) \le S_n(a)+\varepsilon\alpha_n w \quad \text{for all } n \ge n_0 \text{ and } a \in A.$$

We shall denote this by $\quad T_n(A) \le S_n(A)+o(\alpha_n).$

For singleton sets $A = \{a\}$ we shall write $T_n(a) \le S_n(a)+o(\alpha_n)$.

Correspondingly, we shall say $T_n(A) \le S_n(A)+O(\alpha_n)$ if and only if for all $w \in W'$ there are $\varepsilon > 0$ and $n_0 \in N$ such that

$$T_n(a) \le S_n(a)+\varepsilon\alpha_n w \quad \text{for all } n \ge n_0 \text{ and } a \in A$$

If the elements of A are bounded, i.e. their negatives may be adjoined to Q, we denote by $-A = \{-a \mid a \in A\}$.

Note that $\quad S_n(A) \le T_n(A)+o(\alpha_n)$ is equivalent to $T_n(-A) \le S_n(-A)+o(\alpha_n)$,

and that $\quad S_n(A) \le T_n(A)+O(\alpha_n)$ is equivalent to $T_n(-A) \le S_n(-A)+O(\alpha_n)$.

2.2 Remarks. (a) Clearly $T_n(A) \le S_n(A)+o(\alpha_n)$ implies $T_n(A) \le S_n(A)+O(\alpha_n)$. Furthermore, it is easy to check that $T_n(A) \le S_n(A)+O(\alpha_n)$ holds if and only if $T_n(A) \le S_n(A)+o(\alpha'_n)$ holds for every sequence (α'_n) of strictly positive numbers converging to zero such that (α'_n/α_n) converges to infinity: The necessity of the condition is obvious. Suppose on the other hand that the latter conditions holds, but that $T_n(A) \le S_n(A)+O(\alpha_n)$ fails. Then there are $w \in W$, a sequence $(a_n)_{n \in N}$ in A, and a subsequence $(S_{m(n)}, T_{m(n)})_{n \in N}$ of the given operators such that $T_{m(n)}(a_n) \not\le S_{m(n)}(a_n)+n\alpha_{m(n)}w$ for all $n \in N$; a clear contradiction to the above, if we choose the sequence (α'_n) such that $\alpha'_{m(n)} \le (n/2)\,\alpha_{m(n)}$. These observations will allow us to use results on order of convergence of "o-type" in order to obtain results of "O-type".

(b) Note that $T_n(A_i) \leq S_n(A_i)+o(\alpha_n^i)$ for $i = 1,...,m$ implies $T_n(A) \leq S_n(A)+O(\alpha_n)$ for $A = A_1 \cup ... \cup A_m$ and $\alpha_n = \max\{\alpha_n^1,...,\alpha_n^m\}$.

(c) If $\overline{conv(A)}$ denotes the closed (with respect to the symmetric topology) convex hull of A then $T_n(A) \leq S_n(A)+o(\alpha_n)$ implies $T_n(\overline{conv(A)}) \leq S_n(\overline{conv(A)})+o(\alpha_n)$: This is obvious for the convex hull $conv(A)$ of A. Now let $a \in \overline{conv(A)}$ and $w \in \mathcal{W}$. Let $v \in \mathcal{V}$ such that for $b,c \in \mathcal{Q}$

$$b \leq c+v \text{ implies both } T_n(b) \leq T_n(c)+w \text{ and } S_n(b) \leq S_n(c)+w \quad \text{for all } n \in \mathbb{N}.$$

For every $m \in \mathbb{N}$, we can find an element $a_m \in conv(A)$ such that

$$a_m \leq a+(\alpha_m/3)v \quad \text{and} \quad a \leq a_m+(\alpha_m/3)v \quad \text{for all } m \in \mathbb{N}.$$

Now we find $n_0 \in \mathbb{N}$ such that

$$T_n(a_m) \leq S_n(a_m)+(\alpha_n/3)w \quad \text{for all } n \geq n_0 \text{ and } m \in \mathbb{N}.$$

But this shows for $n \geq n_0$

$$T_n(a) \leq T_n(a_n)+(\alpha_n/3)w \leq S_n(a_n)+(2\alpha_n/3)w \leq S_n(a)+\alpha_n w.$$

Now let us consider the sequence cones $(\mathcal{Q}^{\mathbb{N}},\mathcal{V})$ and $(\mathcal{P}^{\mathbb{N}},\mathcal{W})$. To each double sequence $(T_n,S_n)_{n\in\mathbb{N}}$ of u-equicontinuous linear operators from $\mathcal{Q}$ into $\mathcal{P}$ we associate a double sequence $(\overline{T}_n,\overline{S}_n)_{n\in\mathbb{N}}$ of equicontinuous linear operators from $\mathcal{Q}^{\mathbb{N}}$ into $\mathcal{P}^{\mathbb{N}}$ defined for all sequences $(a_n)_{n\in\mathbb{N}} = (a_1,a_2,...) \in \mathcal{Q}^{\mathbb{N}}$ by

$$\overline{T}_n(a_1,a_2,...) = (a_1,...,a_{n-1},T_n(a_n),T_{n+1}(a_{n+1}),...)$$

and, similarly,

$$\overline{S}_n(a_1,a_2,...) = (a_1,...,a_{n-1},S_n(a_n),S_{n+1}(a_{n+1}),...).$$

If, for a fixed element $a \in \mathcal{Q}$ we consider the sequence $(\lambda_1 a,\lambda_2 a,...) \in \mathcal{Q}^{\mathbb{N}}$, where $(\lambda_n)_{n\in\mathbb{N}}$ denotes a sequence of strictly positive numbers, then using the eventual preorder defined in IV.1.7 we have

$$(\overline{T}_n(\lambda_1 a,\lambda_2 a,...)) \preceq (\overline{S}_n(\lambda_1 a,\lambda_2 a,...))$$

if and only if for all $w \in \mathcal{W}$ there is some $n_0 \in \mathbb{N}$ such that

$$T_n(\lambda_n a) \leq S_n(\lambda_n a)+w, \text{ i.e. } T_n(a) \leq S_n(a)+(1/\lambda_n)w \quad \text{for all } n \geq n_0.$$

But this means $T_n(a) \leq S_n(a)+o(1/\lambda_n)$ by our above notation for the order of convergence. Thus indeed, we may study qualitative convergence for the operators $(\overline{T}_n,\overline{S}_n)_{n\in\mathbb{N}}$ from $\mathcal{Q}^{\mathbb{N}}$ into $\mathcal{P}^{\mathbb{N}}$ in order to obtain quantitative results for the operators $(T_n,S_n)_{n\in\mathbb{N}}$ from $\mathcal{Q}$ into $\mathcal{P}$.

3. Smoothness of cone-valued functions.

In order to recover and generalize some of the quantitative approximation results (c.f. [18], [35], [15], [36], [37], [38]) we return to the case of continuous cone-valued functions as in Chapter V: Throughout the remainder of this section, let X denote a locally compact Hausdorff space, Ψ a family of weight functions on X and $(\mathcal{Q},\mathcal{V})$ a locally convex cone. Our criterion for the order of convergence in the Nachbin cone $(C\Psi(X,\mathcal{Q}), \mathcal{V}_\Psi)$ will involve two distinct conditions on the cone-valued function $f \in C\Psi(X,\mathcal{Q})$ under consideration. The first one measures its rate of continuity relative to a certain fixed function on $X \times X$. It is modelled after the real-valued case. We recall the definition of a modulus of continuity:

3.1 Modulus functions. A function $\omega : R^+ \to R^+$ is called a *modulus function* if it is monotone, continuous, $\omega(0) = 0$, and if

$$\omega(t_1+t_2) \le \omega(t_1)+\omega(t_2) \quad \text{holds for all } t_1, t_2 \in R^+.$$

We shall say that ω is of *Lipschitz type* if $\omega(t) = t^\alpha$ for some $0 < \alpha \le 1$. This, of course, is the most important case of modulus functions.

3.2 Remark. The following properties of modulus functions are well-known and easy to check:

(a) $\quad \omega(\rho t) \le (\rho+1)\omega(t) \quad$ for all $\rho, t \ge 0$

(b) $\quad s\,\omega(t) \le 2t\,\omega(s) \quad$ for all $0 \le s \le t$.

In particular, part (b) shows that ω is either the zero-function, or $\omega(t) > 0$ holds for all $t > 0$.

Now we proceed to extend the classical notations as in [18] to Nachbin cones of cone-valued functions.

3.3 Δ^ω-continuous cone-valued functions. Let X be a locally compact Hausdorff space and let

$$\Delta : X \times X \to R$$

denote a continuous non-negative real-valued function on $X \times X$ such that $\Delta(x,x) = 0$ for all $x \in X$, and let ω be a modulus function. We shall say that the $\mathcal{Q}$-valued function f on X is *Δ^ω-continuous* if and only if for every $v \in V$ there is a constant $\rho > 0$ such that

$$f(x) \le f(y) + \rho\,\omega(\Delta(x,y))\,v$$

holds for all $x,y \in X$.

3.4 Remarks. (a) Δ^ω-continuity implies the continuity for the function f with respect to the symmetric topology of $\mathcal{Q}$.

(b) Suppose that for all $x,y \in X$ and $0 \le \lambda \le 1$ there is some $z \in X$ such that $\Delta(x,z) \le \lambda\Delta(x,y)$ and $\Delta(z,y) \le (1-\lambda)\Delta(x,y)$. This holds, for example, if X is a convex subset of a normed space, and if we set $\Delta(x,y) = \|x-y\|^s$ for some $s \ge 1$. (Choose $z = (1-\lambda)x+\lambda y$ for the above condition.) For $f \in C_s(X,\mathcal{Q})$ and $v \in V$ we define

$$\omega_{f,v}(t) = \sup_{\substack{x,y \in X \\ \Delta(x,y) \le t}} \inf\{\rho > 0 \mid f(x) \le f(y) + \rho v\} \quad \text{for } t \ge 0.$$

If its values are finite, then $\omega_{f,V}$ is easily checked to be a modulus function. (In particular, if X is compact and connected, the Remark V.1.3 shows that $\omega_{f,V}(t)$ is finite for all $t \ge 0$.) Finally, for a basis V' of the abstract neighborhood system V we set

$$\omega_{f,V'}(t) = \sup_{v \in V'} \omega_{f,v}(t) \quad \text{for } t \ge 0.$$

If $\omega_{f,v'}$ remains finite, then it is a modulus function as well, and the Q-valued function f is seen to be $\Delta^{\omega_{f,v'}}$-continuous.

(c) Typically, Δ will be defined using finitely many continuous functions $f_1,...,f_N$ on X by the formula

$$\Delta(x,y) = \Big(\sum_{i=1}^{N} (f_i(x)-f_i(y))^2\Big)^s \quad \text{for all} \quad (x,y) \in X \times X,$$

for some $s \geq 1/2$. For $s = 1/2$, this is a semi-metric on X, and the modulus function $\omega_{f,v'}$ as introduced in (b) coincides with the classical modulus of continuity. We shall, however, as a matter of convenience, use $s = 1$ in most of our applications. In that case, Δ^{ω}-continuity holds for a function f if we set $\omega(t) = \omega'(\sqrt{t})$ for all $t \geq 0$, where ω' is the classical modulus of continuity.

Our second condition investigates "smoothness" of the values of f relative to a subset of Q:

3.5 C^{ω}-smooth values. Let X be a locally compact Hausdorff space, ω a modulus function, and let C be a subset of the locally convex cone Q, K a subset of Q^*. Let f be a Q-valued function on X. We shall say that the values of f are *C^{ω}-smooth on K* if $\mu(f(x))$ is finite for all $\mu \in K$, $x \in X$, and if for every u-equicontinuous subset K_0 of K, all $v \in V$ and $\varepsilon > 0$ there is a constant $\rho > 0$ such that for all $\mu \in K_0$, $x \in X$ and for all $0 < \alpha \leq 1$ there is an element c in the convex hull of $\rho C \cup \{0\}$ such that

$$f(x) \leq \frac{\omega(\alpha)}{\alpha} c + \omega(\alpha) v \quad \text{and} \quad \mu\left(\frac{\omega(\alpha)}{\alpha} c\right) \leq \mu(f(x)) + \omega(\alpha)\,\varepsilon.$$

3.6 Examples. (a) If there is $\rho > 0$ such that for all $x \in X$ the values $f(x)$ are contained in the closure (with respect to the symmetric topology of Q) of the convex hull of $\rho C \cup \{0\}$, then clearly the values of f are C^{ω}-smooth on every subset K of Q^* for the modulus function $\omega(t) = t$.

(b) If C is a subcone of Q and the values of the function f are C-superharmonic in all elements of K, then the values of f are seen to be C^{ω}-smooth on K for $\omega(t) = t$: For $\mu \in K$, $v \in V$, $\varepsilon > 0$, $\alpha \leq 1$ and $x \in X$, by the Sup-Inf-Theorem III.1.3 we find an element $c \in C$ such that

$$f(x) \leq c + \alpha v \quad \text{and} \quad \mu(c) \leq \mu(f(x)) + \alpha\varepsilon$$

which yields condition 3.5.

(c) If the values of f are C^{ω_1}-smooth on K and ω_2 is a second modulus function such that $\omega_1(t) \leq \omega_2(t)$ for all $0 \leq t \leq 1$, then the values of f are C^{ω_2}-smooth on K as well: This is obvious for $\omega_2 = 0$, whence $\omega_1 = 0$. Otherwise we conclude: If

$$f(x) \leq \frac{\omega_1(\alpha)}{\alpha} c + \omega_1(\alpha) v \quad \text{and} \quad \mu\left(\frac{\omega_1(\alpha)}{\alpha} c\right) \leq \mu(f(x)) + \omega_1(\alpha)\,\varepsilon$$

as in 3.5, then we set $c' = \dfrac{\omega_1(\alpha)}{\omega_2(\alpha)} c \in \mathrm{conv}(\rho C \cup \{0\})$ and immediately check the above condition with c' in place of c and ω_2 in place of ω_1.

(d) For $Q = CConv(R^n)$ with its usual locally convex cone structure, consider the subset of Q

$$C = \{B, \{e_1\}, \{-e_1\}, \ldots, \{e_n\}, \{-e_n\}\},$$

where B denotes the Euclidean unit ball and $e_1, \ldots, e_n$ the unit vectors in R^n. With each vector $\mu \in R^n$ we associate the functional $\overline{\mu} \in Q^*$ defined for $A \in Q$ by

$\overline{\mu}(A) = \max\{\langle \mu, a\rangle \mid a \in A\}$, where $\langle\, ,\rangle$ denotes the Euclidean inner product.

We set $K = \{\overline{\mu} \in Q^* \mid \mu \in R^n\}$. We shall investigate C^s-smoothness on K for Q-valued functions:

First let $A \in Q$ such that $A \subset \lambda B$ for some $\lambda \geq 0$. Then for $0 \neq \mu \in R^n$ we find an element $a \in A$ such that $\overline{\mu}(A) = \mu(a)$. Now for some $\eta > 0$ we set

$$D = a + \eta\left(B - \frac{\mu}{\|\mu\|}\right) \in Q.$$

The set A is contained in $a + \lambda B$ as well as in the half space $\{b \in R^n \mid \langle \mu, b\rangle \leq \langle \mu, a\rangle\}$, whence, as a simple geometric argument shows, in the set $D + \lambda^2/(2\eta)B$ as well. On the other hand we have $\overline{\mu}(A) = \overline{\mu}(D)$.

Now let f be a Q-valued function on X such that $f(x) \subset \beta B$ for some $\beta > 0$ and for all $x \in X$. We shall show that the values of f are C^ω-smooth on K for the modulus function $\omega(t) = t^{1/2}$: Let K_0 be an equicontinuous subset of K, $v = \delta B \in V$, $\varepsilon > 0$, and set

$$\rho = \beta\left(n + \frac{\beta(n+1)}{2\delta}\right).$$

Then for $0 \neq \overline{\mu} \in K_0$, $x \in X$, and $\alpha \leq 1$ we check the following: We choose

$$a \in f(x) \text{ such that } \overline{\mu}\,(f(x)) = \mu(a).$$

By our introductory investigation we know that for the set

$$D = \alpha^{1/2}\left(a + \frac{\beta^2}{2\delta\alpha^{1/2}}\left(B - \frac{\mu}{\|\mu\|}\right)\right)$$

we have $f(x) \subset \alpha^{-1/2}D + \delta B$ and $\overline{\mu}(\alpha f(x)) = \overline{\mu}(\alpha^{-1/2}D)$. Furthermore, a short computation shows that D is contained in the convex hull of $\sigma C \cup \{0\}$ for some

$$\sigma \leq \alpha^{1/2}\left(n\beta + \frac{\beta^2}{2\delta\alpha^{1/2}}(n+1)\right)$$

$$\leq n\beta + (n+1)\frac{\beta^2}{2\delta} \leq \beta\left(n + \frac{\beta(n+1)}{2\delta}\right) = \rho;$$

thus yielding our claim, indeed.

Smoothness with Lipschitz type modulus functions $\omega(t) = t^\alpha$ for $1/2 < \alpha \leq 1$ may be checked if we add conditions on the curvature of the surface of $f(x)$. Note that a similar statement as the above does not hold if we replace B by an arbitrary unit ball of R^n. It may, however, be verified for those which are strictly convex.

(e) For Q, C, K and a Q-valued function f as in (d) we may look for C^{ω}-smoothness of the values of the (formal) negative of f as well. This turns out to be easier to handle: As before, for $v = \delta B \in V$, $\varepsilon > 0$ and an equicontinuous subset K_0 of K set $\rho = n\beta$. Then for $0 \neq \bar{\mu} \in K_0$, $x \in X$ and $\alpha \leq 1$ we find an element $a \in f(x)$ such that $\bar{\mu}(f(x)) = \langle \mu, a \rangle$. Now we set $c = -\{a\}$ which is clearly contained in the convex hull of $\rho C \cup \{0\}$, and we check immediately that

$$-f(x)) \leq c \quad \text{and} \quad \bar{\mu}(c) = -\bar{\mu}(\alpha(x)) = \bar{\mu}(f(x))$$

holds. Thus, the values of $-f$ are seen to be even C^{ω}-smooth on K for $\omega(t) = t$.

4. A criterion for the order of convergence.

The following theorem is our main tool for studying the order of convergence for Korovkin type approximation in Nachbin cones:

4.1 Theorem. *Let (Q,V) be a full locally convex cone and let K be a subset of Q^* such that $\rho K \subset K$ for all $\rho \geq 0$, and the intersection of K with the polar of each neighborhood is closed.*
Let X be a locally compact Hausdorff space, Ψ a family of weight functions on X, and $p \in C\Psi(X)$ a strictly positive function. Let Δ, C be as in 3.3 and 3.5, and let ω_1 and ω_2 be non-zero modulus functions. For every $v \in V$ suppose that the set E_v consisting of the Q-valued functions

$$x \to \Delta(x,x_0)p(x)\,v \quad \text{for } x_0 \in X,$$

is contained in the Nachbin cone $C\Psi(X,Q)$. Furthermore, by C denote the set of Q-valued functions

$$x \to p(x)\,c \quad \text{for } c \in C.$$

Let $(T_n, S_n)_{n \in N}$ be a double sequence of equicontinuous linear operators from $C\Psi(X,Q)$ into a second locally convex cone (P,W) such that $\bigcap_{n \in N} S_n^{*-1}$ is strictly separating for P and $T_n(E_v) \leq S_n(E_v) + O(\alpha_n)$ for all $v \in V$ and $T_n(C) \leq S_n(C) + O(\beta_n)$.
Then for every bounded positive function $f \in C\Psi(X,Q)$ such that f/p is Δ^{ω_1}-continuous, and the values of f/p are C^{ω_2}-smooth on K, this implies

$$T_n(f) \leq S_n(f) + O(\gamma_n)$$

with

$$\gamma_n = \max\{\omega_1(\alpha_n),\ \omega_2(\beta_n)\}.$$

Furthermore, if $q \in C\Psi(X,Q)$ is a bounded function such that q/p is Δ^{ω_1}-continuous and the values of q/p are C^{ω_2}-smooth on K as well, and if we have in addition

$$S_n(q) \leq T_n(q) + O(\delta_n),$$

then

$$T_n(f) \leq S_n(f) + O(\gamma_n)$$

holds with

$$\gamma_n = \max\{\omega_1(\alpha_n),\ \omega_2(\beta_n),\ \delta_n\}$$

for all (not necessarily positive) functions $f \in C\Psi(X,Q)$ as above such that

$$0 \leq f + \rho q \quad \text{for some } \rho \geq 0.$$

Finally, if the modulus function ω_1 is of Lipschitz type, then all of the above statements still hold true if we replace O-type order of convergence by o-type order of convergence in all assumptions and consequences.

Proof. We may assume that all β_n are at most one (and strictly positive by 2.1). As the modulus functions were supposed to be non-zero we have $\omega_2(\beta_n) > 0$ by 3.2(a). Let $f \in C\Psi(X,Q)$ be a positive function as in the theorem. We shall use the observation on o-type order of convergence from the beginning of this section and study the sequence cones $(C\Psi(X,Q)^N, V)$ and (P^N, W) and the associated double sequence $(\overline{T}_n, \overline{S}_n)_{n\in N}$ of equicontinuous linear operators from $C\Psi(X,Q)^N$ into P^N. We shall use Remark 2.2(a) in order to simultaneously deal with O-type order of convergence in the general case and o-type order of convergence for Lipschitz type modulus functions ω_1: Let (γ_n') be a sequence of strictly positive numbers converging to zero such that $\gamma_n' = \gamma_n$ in the latter case, or such that (γ_n'/γ_n) converges to infinity in the general case. We shall prove that $T_n(f) \leq S_n(f) + o(\gamma_n')$.

Next we fix strictly positive sequences (α_n') and (β_n'), both converging to zero, such that $\alpha_n' = \alpha_n$ and $\beta_n' = \beta_n$ for the Lipschitz case, and for the general case we require

$$\left(\frac{\alpha_n'}{\alpha_n}\right) \to +\infty, \qquad \left(\frac{\beta_n'}{\beta_n}\right) \to +\infty, \qquad \left(\frac{\omega_1(\alpha_n')}{\gamma_n'}\right) \to 0 \qquad \text{and} \qquad \omega_1(\alpha_n'),\ \omega_2(\beta_n') \leq \gamma_n' .$$

(Note that such a choice for the α_n' and β_n' is always possible: Set $\lambda_n = \gamma_n'/\gamma_n$ and $\alpha_n' = (\sqrt{\lambda_n} - 1)\alpha_n$ for sufficiently large n. Then 3.2(a) shows

$$\omega_1(\alpha_n') \leq \sqrt{\lambda_n}\,\omega_1(\alpha_n) \leq \sqrt{\lambda_n}\,\gamma_n = \gamma_n'/\sqrt{\lambda_n},$$

indeed.) Now let $(C\Psi(X,Q)^N)_0$ be the subcone of $C\Psi(X,Q)^N$ generated by the sequences

$$\left(\frac{1}{\alpha_n'} p\cdot e_n + \frac{1}{\beta_n'} p\cdot c_n\right)_{n\in N}$$

for fixed $v \in V$ and arbitrary elements $p\cdot e_n \in E_v$ and $p\cdot c_n \in C$. By our assumption we have $(\overline{T}_n((g_n))) \preceq (\overline{S}_n((g_n)))$ for all sequences $(g_n) \in (C\Psi(X,Q)^N)_0$.
We shall use Proposition 1.2 in order to show $(C\Psi(X,Q)^N)_0$-superharmonicity for the sequence $\left(\frac{1}{\gamma_n'} f\right)_{n\in N}$ (because f is positive, this sequence is bounded below, whence contained in $C\Psi(X,Q)^N$) on the subset $(K_{\Psi X})^N$ of the dual of $C\Psi(X,Q)^N$, which by Proposition V.1.6 and Lemma 1.1 is strictly separating for $C\Psi(X,Q)^N$. This will yield our result: Let (μ_{n,x_n}) be an equicontinuous sequence in $K_{\Psi X}$; i.e. (x_n) is a sequence in X, there is a neighborhood $w \in V$, a weight function $\phi \in \Psi$ such that all x_n are in the support of ϕ, and $\mu_n \in \delta w_Q^\circ \cap K$ for all $\delta > \phi(x_n)$. Now let $v \in V$, $\psi \in \Psi$ and $\varepsilon > 0$. For the criterion of Proposition 1.2 we may assume that $v \leq w$ and $\phi \leq \psi$. Let $\eta > 0$ be an upper bound on X for the function ψp (p is contained in $C\Psi(X)$, whence ψp is bounded). We shall proceed to construct a suitable sequence of functions $(g_n) \in (C\Psi(X,Q)^N)_0$ as in 1.2:
By the Δ^s-continuity of f/p there is a constant $\rho > 0$ such that

$$\frac{f(x)}{p(x)} \leq \frac{f(y)}{p(y)} + \rho\,\omega_1(\Delta(x,y))\,v \qquad \text{for all } x,y \in X.$$

For those $n \in N$ such that $\psi(x_n) = 0$ we observe $\mu_n(f(x_n)) = 0$, as f was supposed to be bounded, and we select $c_n = 0$ in the convex hull of $C \cup \{0\}$.
If on the other hand $\psi(x_n) \neq 0$, then we know from the above that

$$\frac{\mu_n}{\psi(x_n)} \in v_\varrho^\circ \cap K, \quad \text{whence the set} \quad K_0 = \left\{ \frac{\mu_n}{\psi(x_n)} \mid \psi(x_n) \neq 0 \right\}$$

is equicontinuous, and we may use our second assumption on f/p with it. As the values of f/p are supposed to be $C\omega_2$--smooth on K there are a second constant $\rho' \geq 1$ and elements d_n in the convex hull of $\rho' C \cup \{0\}$ such that (we set $\alpha = \beta_n'$ in 3.5) for all such $n \in N$ we have

$$\frac{f(x_n)}{p(x_n)} \leq \frac{\omega_2(\beta_n')}{\beta_n'} d_n + \frac{\omega_2(\beta_n')}{2\eta} v$$

and

$$\left(\frac{\mu_n}{\psi(x_n)}\right)\left(\frac{\omega_2(\beta_n')}{\beta_n'} d_n\right) \leq \left(\frac{\mu_n}{\psi(x_n)}\right)\left(\frac{f(x_n)}{p(x_n)}\right) + \frac{\varepsilon\,\omega_2(\beta_n')}{\eta}.$$

Note that this holds now for all $n \in N$, and multiplying the last inequality with $\psi(x_n)p(x_n)$, this yields

$$\mu_{n,x_n}\left(\frac{\omega_2(\beta_n')}{\beta_n'} p \cdot d_n\right) \leq \mu_{n,x_n}(f) + \varepsilon\,\omega_2(\beta_n').$$

Next we set $c_n = \dfrac{\omega_2(\beta_n')}{\gamma_n'} d_n$, which by $\dfrac{\omega_2(\beta_n')}{\gamma_n'} \leq 1$ is contained in the convex hull of $\rho' C \cup \{0\}$ as well, and rewrite the above inequalities as

$$\frac{1}{\gamma_n'} \frac{f(x_n)}{p(x_n)} \leq \frac{1}{\beta_n'} c_n + \frac{1}{2\eta} v$$

and

$$\mu_{n,x_n}\left(\frac{1}{\beta_n'} p \cdot c_n\right) \leq \mu_{n,x_n}\left(\frac{1}{\gamma_n'} f\right) + \varepsilon.$$

Next we choose a constant σ sufficiently large such that $2\rho < \sigma$, and such that one of the following conditions holds:

in the Lipschitz case: $\quad \omega_1\left(\dfrac{\rho}{\sigma} t\right) \leq \dfrac{1}{2\rho\eta}\omega_1(t) \quad$ for all $t > 0$;

in the general case: there is $n_0 \in N$ such that

$$\omega_1\left(\frac{\rho}{\sigma}\alpha_n'\right) \leq \frac{\gamma_n'}{2\rho\eta} \quad \text{for } n = 1,\dots,(n_0-1), \quad \text{and} \quad \omega_1(\alpha_n') \leq \frac{\gamma_n'}{2\rho\eta} \quad \text{for all } n \geq n_0.$$

(Recall that we required $\left(\dfrac{\omega_1(\alpha_n')}{\gamma_n'}\right) \to 0$.) Now we consider the sequence of functions $g_n \in C\Psi(X,Q)$ defined for all $x \in X$ by

$$g_n(x) = \frac{1}{\beta_n'} p(x)\, c_n + \frac{\sigma}{\alpha_n'} \Delta(x,x_n) p(x)\, v\ .$$

Thus $(g_n) \in (C\Psi(X,Q)^N)_0$, and we clearly have

$$\mu_{n,x_n}(g_n) = \mu_{n,x_n}\left(\frac{1}{\beta_n'} p \cdot c_n\right) \leq \mu_{n,x_n}\left(\frac{1}{\gamma_n'} f\right) + \varepsilon, \quad \text{for all } n \in N.$$

Thus, all left to show in order to verify the criterion of Proposition 1.2 is that

$$\frac{1}{\gamma_n'} f \leq g_n + v_\psi, \quad \text{i.e.} \quad \frac{1}{\gamma_n'} \psi(x) f(x) \leq \psi(x) g_n(x) + v \quad \text{for all } x \in X,$$

holds for all $n \in N$ as well. By the above we have

$$\frac{f(x)}{p(x)} \leq \frac{f(x_n)}{p(x_n)} + \rho\, \omega_1(\Delta(x,x_n))\, v,$$

whence

$$\begin{aligned} \frac{1}{\gamma_n'} f(x) &\leq \frac{f(x_n) p(x)}{\gamma_n' p(x_n)} + \frac{\rho\, \omega_1(\Delta(x,x_n))}{\gamma_n'} p(x)\, v \\ &\leq \frac{p(x)}{\beta_n'} c_n + \left(\frac{\rho\, \omega_1(\Delta(x,x_n))}{\gamma_n'} + \frac{1}{2\eta} \right) p(x)\, v . \end{aligned}$$

But this shows that $\frac{1}{\gamma_n'} \psi(x) f(x) \leq \psi(x) g_n(x) + v$ holds, whenever

$$\frac{\rho\, \omega_1(\Delta(x,x_n))}{\gamma_n'} \leq \frac{\sigma \Delta(x,x_n)}{\alpha_n'}, \quad \text{i.e.} \quad \frac{\omega_1(\Delta(x,x_n))}{\Delta(x,x_n)} \leq \frac{\sigma}{\rho} \frac{\gamma_n'}{\alpha_n'}.$$

Thus, we have to investigate the case

$$\frac{\omega_1(\Delta(x,x_n))}{\Delta(x,x_n)} \geq \frac{\sigma}{\rho} \frac{\gamma_n'}{\alpha_n'} \geq \frac{\sigma}{\rho} \frac{\omega_1(\alpha_n')}{\alpha_n'} :$$

Firstly, we claim that the above inequality implies that $\Delta(x,x_n) \leq \alpha_n'$; otherwise, by 3.2 (b), we would infer

$$\frac{\omega_1(\Delta(x,x_n))}{\Delta(x,x_n)} \leq 2 \frac{\omega_1(\alpha_n')}{\alpha_n'} ,$$

a contradiction to the above, as we required $\frac{\sigma}{\rho} > 2$.

Secondly, now solving the above inequality for $\Delta(x,x_n)$ and considering that $\omega_1(\Delta(x,x_n)) \leq \omega_1(\alpha_n')$, we even infer

$$\Delta(x,x_n) \leq \frac{\rho}{\sigma} \alpha_n'.$$

Now, in the Lipschitz case, using our second requirement for σ, we have

$$\frac{\rho\, \omega_1(\Delta(x,x_n))}{\gamma_n'} \leq \frac{\rho\, \omega_1\!\left(\frac{\rho}{\sigma} \alpha_n'\right)}{\gamma_n'} \leq \frac{\omega_1(\alpha_n')}{2\eta \gamma_n'} \leq \frac{1}{2\eta}.$$

In the general case, however, we conclude

$$\frac{\rho\, \omega_1(\Delta(x,x_n))}{\gamma_n'} \leq \frac{\rho\, \omega_1(\alpha_n')}{\gamma_n'} \leq \frac{1}{2\eta} \quad \text{for } n \geq n_0,$$

and by our choice of σ

$$\frac{\rho\, \omega_1(\Delta(x,x_n))}{\gamma_n'} \leq \frac{\rho\, \omega_1\!\left(\frac{\rho}{\sigma} \alpha_n'\right)}{\gamma_n'} \leq \frac{1}{2\eta} \quad \text{for } n = 1, \ldots, (n_0 - 1)$$

as well. Thus, indeed, we have verified in all cases and for all $x \in X$

$$\begin{aligned} \frac{1}{\gamma_n} \psi(x) f(x) &\leq \frac{1}{\gamma_n^{1/t}} \psi(x) p(x) c_n + \left(\frac{1}{2\eta} + \frac{1}{2\eta}\right) \psi(x) p(x)\, v \\ &\leq \psi(x) g_n(x) + v\,, \end{aligned}$$

as desired.

For the last part of our theorem, the arguments in the general and in the Lipschitz case are analogous. We shall proceed only for the general case: Let $f \in C\Psi(X,Q)$ be as before, but not necessarily positive. For the bounded function $q \in C\Psi(X,Q)$, let q/p be Δ^{ω_1}-continuous, its values C^{ω_2}-smooth on K, and suppose $S_n(q) \leq T_n(q) + O(\delta_n)$. By our assumption there is $\rho \geq 0$ such that the function $F = f + \rho q$ is positive. Then F/p is seen to be Δ^{ω_1}-continuous and its values are C^{ω_2}-smooth on K as well. Thus, the first part of our theorem applies to the function F: For $w \in W$ there are $\sigma > 0$ and $n_0 \in N$ such that both

$$T_n(F) = T_n(f) + T_n(\rho q) \leq S_n(F) + \sigma \gamma_n w = S_n(f) + S_n(\rho q) + \sigma \gamma_n w$$

and

$$S_n(\rho q) \leq T_n(\rho q) + \sigma \gamma_n w$$

hold for all $n \geq n_0$. But this implies

$$T_n(f) + T_n(\rho q) \leq S_n(f) + T_n(\rho q) + 2\sigma \gamma_n w.$$

Finally, $T_n(\rho q)$ is bounded in P, as p is bounded in $C\Psi(X,Q)$ and thus may be cancelled (c.f. Ch I, Proposition 4.4) in the last inequality, yielding

$$T_n(f) \leq S_n(f) + 2\sigma \gamma_n w \leq S_n(f) + 3\sigma \gamma_n w$$

This shows $T_n(f) \leq S_n(f) + O(\gamma_n)$, and completes our proof.

4.2 Remarks. (a) If the negatives of all neighborhoods $v \in V$ are contained in Q as well, and if Δ is as in Remark 3.4 (c) with some $s \in N$, i.e.

$$\Delta(x,y) = \Big(\sum_{i=1}^{N} (f_i(x) - f_i(y))^2\Big)^s \quad \text{for functions} \quad f_1, \dots, f_N \in C\Psi(X),$$

then the first condition on the double sequence $(T_n,S_n)_{n \in N}$ of operators in the preceding theorem may be simplified, because the sets E_v then are generated by the finitely many functions

$$x \to p(x)\, v, \quad x \to -p(x) f_i(x)^{(2k-1)}\, v \quad \text{and} \quad x \to p(x) f_i(x)^{2k}\, v,$$

for $i = 1, \dots, N$ and $k = 1, \dots, s$. Thus we only have to check the order of convergence on those functions; more precisely:

The condition $T_n(E_v) \leq S_n(E_v) + O(\alpha_n)$ from Theorem 4.1 translates into

$$T_n(p \cdot v) \leq S_n(p \cdot v) + O(\alpha_n), \qquad T_n(p f_i^{2k} \cdot v) \leq S_n(p f_i^{2K} \cdot v) + O(\alpha_n)$$

and

$$S_n(p f_i^{(2k-1)} \cdot v) \leq T_n(p f_i^{(2k-1)} \cdot v) + O(\alpha_n)$$

for $i = 1, \dots, N$ and $k = 1, \dots, s$.

(b) Clearly, Theorem 4.1 provides a sufficient criterion for the order of convergence if the operators under consideration are defined only on a subcone of $C\Psi(X,Q)$. All functions mentioned in the criterion, however, have to be contained in this subcone as well.

In the following final corollaries to this theorem we shall restrict ourselves to sequences of operators converging towards the identity. It is, however, obvious how the upcoming results might be formulated for more general operators S_n. Throughout the remainder of this section, let us adopt the following notations:

(Q,V) is a full locally convex cone,
X a locally compact Hausdorff space,
Ψ a family of weight functions on X,
$p \in C\Psi(X)$ a strictly positive function
and $\Delta : X \times X \to R$ is defined using $f_1,...,f_N \in C\Psi(X)$, as

$$\Delta(x,y) = \sum_{i=1}^{N} (f_i(x)-f_i(y))^2.$$

Our first corollary recovers de Vore's [18] classical result for the order of convergence of positive linear operators on $C(X)$: We set $Q = R$ with its canonical neighborhood system $V = \{\rho > 0\}$. The dual cone R^* may be identified with $R^+ = \{\alpha \geq 0\}$, and for our theorem we choose $K = R^+$. Furthermore, if we set $C = \{-1,+1\} \subset R$, then for every function $f \in C\Psi(X)$ such that $|f| \leq \lambda p$ for some $\lambda > 0$, the function f/p has C^ω-smooth values on K for $\omega(t) = t$, and the last assumption of Theorem 4.1 holds with this function $q = p$ as well. Summarizing, this yields:

4.3 Corollary. *Let* $(T_n)_{n \in N}$ *be an equicontinuous sequence of positive linear operators on* $C\Psi(X)$. *If*

$$T_n(p) \leq p+O(\alpha_n), \quad T_n(pf_i^2) \leq pf_i^2+O(\alpha_n), \quad pf_i \leq T_n(pf_i)+O(\alpha_n)$$

and $\qquad p \leq T_n(p)+O(\beta_n) \quad$ *for* $i = 1,..., N$,

then $\qquad T_n(f) \leq f+O(\max\{\omega(\alpha_n), \beta_n\})$

for every function $f \in C\Psi(X)$ *such that* f/p *is* Δ^ω-*continuous and* $|f| \leq \lambda p$ *for some* $\lambda > 0$. *If the modulus function* ω *is of Lipschitz type, then an analogous statement holds for o-type order of convergence as well.*

Next we turn to contractive linear operators on the space $C_{\mathbb{C}}(X)$ of complex-valued continuous functions on a compact Hausdorff space X. We consider the usual sup-norm topology on $C_{\mathbb{C}}(X)$, i.e. $\Psi = \{1\}$ and $p = 1$. We prepare the upcoming corollary with a technical lemma. As usual, by $C(X)$ we denote the subset of real-valued continuous functions in $C_{\mathbb{C}}(X)$, by $C(X)^+$ the positive cone in $C(X)$.

4.4 Lemma. *Let* T *be a (complex) linear operator on* $C_{\mathbb{C}}(X)$ *and* $0 \leq \varepsilon < 1$ *such that both* $\|T\| \leq (1+\varepsilon)$ *and* $\|T(1)-1\| < \varepsilon$. *Then the following holds:*

(i) *For every* $f \in C(X)$ *we have* $\|\mathrm{Im}(T(f))\| \leq 2\varepsilon^{1/2}\|f\|$.

(ii) *For every* $f \in C(X)^+$ *we have* $T(f) = p+r$, *for some* $p \in C(X)^+$ *and* $r \in C_{\mathbb{C}}(X)$ *such that* $\|r\| \leq 3\varepsilon^{1/2}\|f\|$.

(iii) *For every* $f \in C_{\mathbb{C}}(X)$ *we have* $\|\mathrm{Re}(T(f)) - T(\mathrm{Re}(f))\| \leq 4\varepsilon^{1/2}\|f\|$.

Proof. Note that we may restrict ourselves to the case $\|f\| \leq 1$, as the general case follows from this by multiplication with a suitable scalar.

(i) Let $f \in C(X)$ such that $\|f\| \leq 1$ and set $T(f) = g+ih$ for $g,h \in C(X)$. For $h = 0$ our statement is obvious. Otherwise, let $x \in X$ such that $|h(x)| = \|h\| = \|\mathrm{Im}(T(f))\|$. Then using the assumptions on T and setting $\rho = 1/\|h\|$, this shows for every $\lambda \geq 0$

$$\|T(f+i\lambda/\rho)\| = \|g+i(h+\lambda/\rho T(1)\| = \|g+i(h+\lambda/\rho)+\lambda/\rho(T(1)-1)\|$$
$$\leq (1+\varepsilon)\|f+i\lambda/\rho\| \leq (1+\varepsilon)\sqrt{1+\lambda^2/\rho^2}\,.$$

Using $\|T(1)-1\| \leq \varepsilon$, this yields

$$\|g+i(h+\lambda/\rho)\| \leq (1+\varepsilon)\sqrt{1+\lambda^2/\rho^2} + \varepsilon\lambda/\rho$$

and in particular (recall $|h(x)| = 1/\rho$)

$$(1/\rho)(1+\lambda) \leq |g(x)+i(h(x)+\lambda/\rho)| \leq (1+\varepsilon)\sqrt{1+\lambda^2/\rho^2} + \varepsilon\lambda/\rho;$$

whence
$$(1+\lambda) \leq (1+\varepsilon)\sqrt{\rho^2+\lambda^2} + \varepsilon\lambda$$
and
$$1+(1-\varepsilon)\lambda \leq (1+\varepsilon)\sqrt{\rho^2+\lambda^2}\,.$$

Squaring both sides and sorting the coefficients, this yields after a short computation that the inequality

$$4\varepsilon\,\lambda^2 + 2(\varepsilon-1)\,\lambda + ((1+\varepsilon)^2\rho^2-1) \geq 0$$

must hold true for all $\lambda \geq 0$. An easy computation shows that the above quadratic polynomial in λ attains its minimum value of $((1+\varepsilon)^2\rho^2-1)-(1-\varepsilon)^2/4\varepsilon$ in $\lambda = (1-\varepsilon)/4\varepsilon > 0$. This leads to

$$(1-\varepsilon)^2 \leq 4\varepsilon((1+\varepsilon)^2\rho^2-1)$$

and after a few more steps to $1 \leq 4\varepsilon\rho^2$. But the latter yields our claim.

(ii) Now let $f \in C(X)^+$ such that $\|f\| \leq 1$ and set $F = f-1/2$. Then we have $\|F\| \leq 1/2$ and $\|T(F)\| \leq (1+\varepsilon)/2$. From (i) we know

$$T(F) = g+ih \quad \text{such that} \quad \|h\| \leq \varepsilon^{1/2}, \quad \text{whence} \quad \|g\| \leq 1/2+\varepsilon^{1/2}.$$

Furthermore, from

$$T(f) = T(F)+T(1/2) = (g+1/2+\varepsilon^{1/2}) + (ih - \varepsilon^{1/2}+1/2(T(1)-1))$$

we see, indeed, that the function $p = g+1/2+\varepsilon^{1/2}$ is positive and that

$$\|r\| = \|ih - \varepsilon^{1/2}+1/2(T(1)-1)\| \leq \varepsilon^{1/2}+\varepsilon^{1/2}+(1/2)\varepsilon \leq 3\varepsilon^{1/2}.$$

(iii) Now let $f = g+ih \in C_{\mathbb{C}}(X)$. Then

$$\mathrm{Re}(T(f)) = \mathrm{Re}(T(g))-\mathrm{Im}(T(h)) = (T(g)-i\,\mathrm{Im}(T(g)))-\mathrm{Im}(T(h));$$

thus by (i)

$$\|\mathrm{Re}(T(f))-T(\mathrm{Re}(f))\| = \|\mathrm{Re}(T(f))-T(g)\| \leq \|\mathrm{Im}(T(g))\|+\|\mathrm{Im}(T(h))\| \leq 4\varepsilon^{1/2}\|f\|$$

which completes our proof.

Now we may formulate our second corollary. Continuity of functions will refer to the canonical topology of $\mathbb{C}$. Note that for a sequence of operators (T_n) on $C_{\mathbb{C}}(X)$ convergence on a function now means convergence with respect to the usual topology, i.e. $T_n(f) \leq f+O(\alpha_n)$ and $T_n(-f) \leq -f+O(\alpha_n)$ coincide.

4.5 Corollary. *Let X be a compact Hausdorff space and $(T_n)_{n\in N}$ a sequence of (complex) linear operators on $C_{\mathbb{C}}(X)$. If*

$$\|T_n\| \le 1+O((\alpha_n)^2),\quad \|T_n(1)-1\| \le O((\alpha_n)^2),$$

$$\|T_n(f_i)-f_i\| \le O(\alpha_n) \quad \textit{and} \quad \|T_n(f_i^2)-f_i^2\| \le O(\alpha_n) \quad \textit{for}\ \ i=1,\dots,N,$$

then

$$\|T_n(f)-f\| \le O(\omega(\alpha_n))$$

for every Δ^{ω}-continuous function $f\in C_{\mathbb{C}}(X)$.
If the modulus function ω is of Lipschitz type, then an analogous statement holds for o-type order of convergence as well.

Proof. By B we denote the unit ball and by Γ the unit circle in $\mathbb{C}$. As $\mathbb{C}$ itself, endowed with the neighborhood system $V=\{\rho B \mid \rho>0\}$, is not a full cone we we have to consider the full locally convex cone $Q=\{c+\rho B \mid c\in\mathbb{C},\ \rho\ge 0\}$ instead. Q is ordered by inclusion. Moreover, every continuous Q-valued function $F\in C_s(X,Q)$ may be expressed using a complex-valued function f and a positive real-valued function g on X, as

$$F(x)=f(x)+g(x)B \ \text{ for all } \ x\in X.$$

Clearly, f and g are uniquely determined and we observe that for $x,y\in X$ and $\rho B\in V$ we have

$$F(x)\le F(y)+\rho B \ \text{ if and only if } \ \|f(x)-g(x)\| \le g(y)-g(x).$$

But this shows that for continuous F both functions f and g have to be continuous as well. Turning to a continuous linear operator T on $C_{\mathbb{C}}(X)$ we observe that in this case there is no straightforward extension to an operator on $C_s(X,Q)$. However, we now have to turn to the range cone $\mathbf{Conv}(C_s(X,Q))$ in order to define the corresponding operator $\overline{T}: C_s(X,Q)\to \mathbf{Conv}(C_s(X,Q))$ by

$$F\to \{f\in C_s(X,Q)\mid f\le F\} \ :\ C_s(X,Q)\to \mathbf{Conv}(C_s(X,Q))$$

which is clearly u-continuous and coincides on $C_{\mathbb{C}}(X)$ with the given operator T. A suitable subset K of Q^* as in Theorem 4.1 is given by $K=\{\gamma\oplus|\gamma| \mid \gamma\in\mathbb{C}\}$, where the elements of K operate on Q as

$$\gamma\oplus|\gamma|\,(c+\rho B)=\mathrm{Re}(\gamma c)+\rho|\gamma|\,.$$

Moreover, for the extended identity operator Id in place of the operators S_α it is immediately checked that the respective condition from Theorem 4.1 holds with this subset K. Now we may directly apply Theorem 4.1 choosing the one-function for p: Consider the set of real-valued functions E in $C(X)$

$$x\to (f_i(x)-f_i(x_0))^2 \quad \text{for all} \ \ x_0\in X,\ \ i=1,\dots,N$$

and the corresponding set E_B of functions in $C_s(X,Q)$

$$x\to (f_i(x)-f_i(x_0))^2B \quad \text{for all} \ \ x_0\in X,\ \ i=1,\dots,N.$$

We have to check the order of equiconvergence of our extended operators $\overline{T_n}$ on this latter set of functions: For $e\in E$ we denote by $e\cdot B$ the corresponding Q-valued function in E_B. Firstly, from our assumptions on the operators T_n, we we find $\rho>0$ and $n_0\in N$ such that

$$\|T_n\|\le 1+(\rho\alpha_n)^2 \quad \text{and} \quad \|T_n(e)-e\| < \rho\alpha_n \quad \text{for all} \ \ n\ge n_0 \ \text{ and } \ e\in E.$$

Now let $e \in E$ and let $f \in C_{\mathbb{C}}(X)$ such that

$$f \leq e \cdot B, \quad \text{i.e.} \quad f(x) \leq e(x)B \quad \text{for all} \quad x \in X, \quad \text{i.e.} \quad |f| \leq e.$$

Then for all $\gamma \in \Gamma$, using Lemma 4.4 (iii) this shows

$$|\mathrm{Re}(\gamma T_n(f))| = |\mathrm{Re}(T_n(\gamma f))| \leq |T_n(\mathrm{Re}(\gamma f))| + 4\rho\alpha_n.$$

Using 4.4 (ii) we proceed

$$|T_n(\mathrm{Re}(\gamma f))| = |T_n(\mathrm{Re}(\gamma f)-e)| + |T_n(e)-e| \leq e - p + 3\rho\alpha_n + \rho\alpha_n,$$

where p is a positive real-valued function. Combining the above yields

$$|\mathrm{Re}(\gamma T_n(f))| \leq e + 8\rho\alpha_n$$

for all $\gamma \in \Gamma$, whence

$$|T_n(f))| \leq e + 8\rho\alpha_n; \quad \text{i.e.} \quad T_n(f) \leq e \cdot B + 8\rho\alpha_n B, \quad \text{whenever} \quad f \leq eB.$$

But the latter shows for the functions $eB \in E_B$ and the extended operators $\overline{T}_n$

$$\overline{T}_n(e \cdot B) \leq e \cdot B + 8\rho\alpha_n B, \quad \text{i.e.} \quad \overline{T}_n(E_B) \leq E_B + O(\alpha_n).$$

Thus, the first assumption in Theorem 4.1 has been verified. For the second part we choose the subset $C = \{\gamma + B \mid \gamma \in B\}$ of Q and denote by C the set of constant functions with values in C. We easily check for every function in $C_s(X,Q)$ that its values are C^{ω}-smooth for $\omega(t) = t$. Furthermore, by our assumption we have

$$\overline{T}_n(C) \leq C + O((\alpha_n)^2).$$

Thus, Theorem 4.1 applies to all positive Δ^{ω}-continuous functions in $C_s(X,Q)$ with $\gamma_n = \omega(\alpha_n)$. Finally, let $f \in C_{\mathbb{C}}(X)$ be any Δ^{ω}-continuous single-valued function and $\rho > 0$ such that the function $f + \rho B$ is positive in $C_s(X,Q)$. Of course, Δ^{ω}-continuity for f and for $f + \rho B$ coincide. The above shows that we find $\sigma > 0$ and $n_0 \in N$ such that for all $n \geq n_0$

$$T_n(f + \rho B) \leq f + \rho B + \sigma\omega(\alpha_n)B$$

holds. For any $\gamma \in \rho\Gamma$ this implies

$$T_n(f+\gamma) = T_n(f) + T_n(\gamma) = T_n(f) + \gamma + (T_n(\gamma) - \gamma) \leq f + (\rho + \sigma\omega(\alpha_n))B,$$

i.e. $\quad |(T_n(f)(x) - f(x)) + \gamma| \leq \rho + \sigma\omega(\alpha_n) + \|(T_n(\gamma) - \gamma)\| = \rho + \sigma\omega(\alpha_n) + \rho\|(T_n(1) - 1)\|$

for all such γ and $x \in X$. But the latter shows

$$|(T_n(f)(x) - f(x))| \leq \sigma\omega(\alpha_n) + \rho\|(T_n(1) - 1)\|,$$

whence, indeed, $\quad \|T_n(f) - f\| \leq O(\omega(\alpha_n)).$

Next we turn to monotone linear operators on spaces of continuous set-valued functions. For our first result, let Q be the locally convex cone $CConv(R^m)$ of the non-empty compact convex subsets of R^m, ordered by inclusion and endowed with the neighborhood system $V = \{\rho B \mid \rho > 0\}$ where B denotes the Euclidean unit ball of R^m. As Q is a full cone we may directly apply Theorem 4.1: Let X, Ψ, Δ and p be as before. Consider the subset C of $C\Psi(X, CConv(R^m))$ as in Example 3.6(d). The values of f/p for every function $f \in C\Psi(X, CConv(R^m))$ such that $f(x) \subset \lambda p(x)B$ for some $\lambda > 0$ and all $x \in X$ were seen to be C^{ω}-smooth on $K = \{\overline{\mu} \in Q^* \mid \mu \in R^m\}$ with $\omega(t) = t^{1/2}$ in that example. For the last statement of Theorem 4.1 we may use the function $x \to p(x)B$. Finally, we may also consider the formal negatives of the functions in $C\Psi(X, CConv(R^m))$ which have values in $-CConv(R^m)$. (Formally, we have to introduce the locally convex cone $C\Psi(X,CConv(R^m)\text{-}CConv(R^m))$). Clearly, $-f/p$ is Δ^{ω}-continuous, whenever this holds for the function f/p. Moreover, its values were seen to be even C^{ω}-smooth on K for $\omega(t) = t$ in Example 3.6 (e).

Now Theorem 4.1 yields immediately:

4.6 Corollary. *Let* $(T_n)_{n\in N}$ *be an equicontinuous sequence of monotone linear operators on* $C\Psi(X,CConv(R^m))$. *If*

$$pf_i\cdot B \le T_n(pf_i\cdot B)+O(\alpha_n), \qquad T_n(pf_i^2\cdot B) \le pf_i^2\cdot B+O(\alpha_n),$$
$$T_n(p\cdot B) \le p\cdot B+O(\beta_n), \qquad p\cdot B \le T_n(p\cdot B)+O(\delta_n)$$

and
$$T_n(p\cdot\{\pm e_j\}) \le p\cdot\{\pm e_j\}+O(\varepsilon_n)$$

for $i = 1,\dots, N$ *and* $j = 1,\dots, m$, *then*

$$T_n(f) \le f+O(\gamma_n) \quad and \quad f \le T_n(f)+O(\eta_n)$$

for every function $f \in C\Psi(X,CConv(R^n))$ *such that* f/p *is* Δ^{ω}*-continuous and* $f(x) \subset \lambda p(x)B$ *for some* $\lambda > 0$, *with*

$$\gamma_n = \max\{\omega(\alpha_n), \omega(\beta_n), (\beta_n)^{1/2}, (\varepsilon_n)^{1/2}, \delta_n\} \quad and \quad \eta_n = \{\omega(\alpha_n), \omega(\beta_n), \beta_n, \varepsilon_n, \delta_n\}.$$

If the modulus function ω *is of Lipschitz type, then an analogous statement holds for o-type order of convergence as well.*

4.7 Example. For $X = [0,1]$, the weight functions $\Psi = \{1\}$ induce the topology of uniform convergence on $[0,1]$. We set $p = 1$ and and define Δ with the single function $f_1(x) = x$, i.e. $\Delta^{1/2}$ defines the usual metric of the interval $[0,1]$. As an example for a sequence of monotone linear operators on $C\Psi([0,1],CConv(R^n))$ we consider the *Bernstein* operators BS_n defined for all set-valued functions $f \in C\Psi([0.1],CConv(R^n))$ and $x \in [0,1]$ by

$$BS_n(f)(x) = \sum_{\kappa=0}^{n} \binom{n}{\kappa} f\left(\frac{\kappa}{n}\right) x^{\kappa}(1-x)^{n-\kappa}.$$

It is obvious that the operators BS_n all are monotone and u-continuous. Furthermore, an easy computation shows that for all $n \in N$ we have

$$BS_n(K) = K$$

for every constant set-valued function $K \in C\Psi(\{0.1\],CConv(R^n))$ and for the special functions we have

$$BS_n(x\cdot B) = x\cdot B \quad \text{and} \quad BS_n(x^2\cdot B) = x^2\cdot B + \frac{1}{n}(x-x^2)\cdot B\ .$$

Thus, we have checked

$$BS_n(x^2\cdot B) \le x^2\cdot B+O\left(\frac{1}{n}\right)$$

and may apply Corollary 4.6 with those α_n and any choice for β_n, δ_n and ε_n. This yields

$$BS_n(f) \le f + O\left(\omega\left(\frac{1}{n}\right)\right) \quad \text{and} \quad f \le BS_n(f) + O\left(\omega\left(\frac{1}{n}\right)\right)$$

for every Δ^{ω}-continuous set-valued function $f \in C\Psi([0,1], CConv(R^n))$.

4.8 Example. For $X = R$, $\Psi = \{1\}$, i.e. $C\Psi(R) = C_0(R)$ we shall give an example for the approximation of set-valued functions in the two-dimensional case: First consider the operators P_n on $CConv(R^2)$ which associate with every non-empty compact convex subset A of R^2 the subset $P_n(A)$ which we obtain as follows: Let x_j be the vectors in R^2

$$x_j = (\cos(2j\pi/n), \sin(2j\pi/n)), \quad j = 1,\dots,n$$

and set

$$P_n(A) = \{b \in R^2 \mid \langle b,x_i\rangle \le \max\{\langle a,x_i\rangle \mid a \in A\} \text{ for all } i = 1,\dots,n\}.$$

The operators P_n are seen to be linear and monotone on $CConv(R^2)$. Their values are polygonal areas with at most n vertices and segments orthogonal to the vectors x_j. An easy computation shows for the Euclidean unit ball B

$$B \subset P_n(B) \subset B + \frac{2}{n^2} B \quad \text{for sufficiently large } n \in N.$$

For the singleton sets $\{\pm e_j\}$, $j = 1, 2$, we observe that $P_n(\{\pm e_j\}) = \{\pm e_j\}$ for all $n \ge 3$. Next we select the function $p \in C_0(R)$ as $p(x) = e^{-x^2}$, and again define Δ using the single function $f_1(x) = x$.

Now we consider the operators T_n on $C_0(R, CConv(R^2))$ defined for $f \in C_0(R, CConv(R^2))$ and $x \in R$ in the following way: For every integer number κ let $x_\kappa = \kappa/n$, and for $x_\kappa \le x < x_{\kappa+1}$ set $\lambda = n(x - x_\kappa)$ and

$$T_n(f)(x) = (1-\lambda)\, e^{(x_\kappa)^2 - x^2} P_n(f(x_\kappa)) + \lambda\, e^{(x_{\kappa+1})^2 - x^2} P_n(f(x_{\kappa+1})).$$

Clearly, those operators are linear, monotone and u-continuous. For the functions e_1, e_2 and e_3 such that

$$e_1(x) = e^{-x^2} B, \quad e_2(x) = x e^{-x^2} B, \quad \text{and} \quad e_3(x) = x^2 e^{-x^2} B$$

we check

$$T_n(e_1)(x) = e^{-x^2} P_n(B), \quad T_n(e_2)(x) = x e^{-x^2} P_n(B)$$

and

$$T_n(e_3)(x) = \left(x^2 + \left(\frac{\lambda - \lambda^2}{n^2}\right)\right) e^{-x^2} P_n(B).$$

Now our above estimate for the operators P_n shows that the assumptions of Corollary 4.6 hold for this sequence of operators with $\alpha_n = \beta_n = 1/n^2$, and any choice of the δ_n. Finally, as $T_n(e^{-x^2} \cdot \{\pm e_j\}) = e^{-x^2} \cdot \{\pm e_j\}$, for $j = 1, 2$ and $n \ge 3$, we may arbitrarily choose the ε_n as well.

Now Corollary 4.6 yields

$$T_n(f) \le f + O\left(\omega\left(\frac{1}{n}\right)\right) \quad \text{and} \quad f \le T_n(f) + O\left(\omega\left(\frac{1}{n}\right)\right)$$

for every function $f \in C_0(R, CConv(R^2))$ such that $e^{x^2} f$ is Δ^{ω}-continuous.

Note that in the above example the order of convergence is $(1/n^2)$ for the test set of functions, but for a general function $f \in C_0(R, CConv(R^2))$ we may conclude only convergence of order $\omega(1/n)$, due to only $C^{\omega'}$-smoothness for $\omega'(t) = t^{1/2}$ of the values of f. This result, however, is optimal for general functions, as the following may demonstrate: Let $U = [0, 1] \in CConv(R^2)$. It is easy to check that

$$U \subset P_n(U) \subset U + \frac{4}{n}B$$

holds for sufficiently large n, and any stronger rate of convergence is seen to fail. Thus for the function $f \in C_0(R, CConv(R^2))$, defined as

$$f(x) = e^{-x^2}\, U \quad \text{for all} \quad x \in R,$$

we clearly have Δ^{ω}-continuity with $\omega(t) = t$ for $e^{x^2}f$. And for $T_n(f)$ we check by the above that the order of convergence is only $(1/n)$, indeed.

A similar construction of operators P_n as in the above example may be used in the spaces R^m.

There are various ways to generalize and modify Corollary 4.6 in order to suit the case of set-valued functions for infinite dimensional vector spaces. They all suffer from the fact that it is hard to identify a suitable subset K of the dual cone as required in Theorem 4.1. Moreover, having fixed a family C of convex subsets of the vector space, it will be a tough task to establish C^{ω}-smoothness for the values of a given set-valued function. (This will already be difficult, if we replace the Euclidean unit ball B of R^m in 4.6 by any other one.) Thus, we restrict ourselves to giving a manageable, however somewhat crude, result for normed spaces in the following Corollary:

Let $(E, \|\ \|)$ be a normed space with unit ball B. As in V.3.13 we denote by $\overline{BConv(E)}$ the locally convex cone of all non-empty bounded and closed subsets of E. For the set K in Theorem 4.1 we choose all of $\overline{BConv(E)}^*$, and $C = \{K \in \overline{BConv(E)} \mid K \subset B\}$. The latter choice trivially guarantees C^{ω}-smoothness for the values of every function $f \in C\Psi(X, \overline{BConv(E)})$ with the modulus function $\omega(t) = t$ (c.f. Example 3.6 (a)). As before, we denote by $\mathcal{C} = \{p \cdot K \mid K \in C\}$. For the last part of Theorem 4.1 we use the function $p \cdot B$.

Using this, the following is immediate:

4.9 Corollary. *Let $(E, \|\ \|)$ be a normed vector space with unit ball B. Let $(T_n)_{n \in N}$ be an equicontinuous sequence of monotone linear operators on $C\Psi(X, \overline{BConv(E)})$. If*

$$T_n(p \cdot B) \le p \cdot B + O(\alpha_n), \quad pf_i \cdot B \le T_n(pf_i \cdot B) + O(\alpha_n), \quad T_n(pf_i^2 \cdot B) \le pf_i^2 \cdot B + O(\alpha_n),$$

$$p \cdot B \le T_n(p \cdot B) + O(\beta_n) \quad \text{and} \quad T_n(\mathcal{C}) \le \mathcal{C} + O(\beta_n) \quad \text{for } i = 1, \dots, N,$$

then $$T_n(f) \le f + O(\max\{\omega(\alpha_n), \beta_n\}) \quad \text{and} \quad f \le T_n(f) + O(\max\{\omega(\alpha_n), \beta_n\})$$

for every function $f \in C\Psi(X, \overline{BConv(E)})$ such that f/p is Δ^{ω}-continuous and $f(x) \subset \lambda p(x)B$ for some $\lambda > 0$.
If the modulus function ω is of Lipschitz type, then an analogous statement holds for o-type order of convergence as well.

To give an example for this last corollary, as in Example 4.7, let $X = [0,1]$, $\Psi = \{1\}$, $p = 1$, and consider the Bernstein operators BS_n on $C\Psi(X, \overline{BConv(E)})$ defined by the same formula as in 4.7. They were seen to keep the constant set-valued functions invariant and convergent of order $O\left(\frac{1}{n}\right)$ on the remaining test function. Thus, as in 4.7, now using Corollary 4.9, we conclude again that

$$BS_n(f) \le f + O\left(\omega\left(\frac{1}{n}\right)\right) \quad \text{and} \quad f \le BS_n(f) + O\left(\omega\left(\frac{1}{n}\right)\right)$$

holds for every Δ^{ω}-continuous set-valued function $f \in C\Psi([0,1], \overline{BConv(E)})$.

For our final corollary we return to the approximation of stochastic processes as in our Example V.2.6. The quantitative result which we shall recover is due to M. Weba [61]. Let X, Ψ, p and Δ be as before. As in V.2.6, let (Ω, A, σ) be a probability space and L^1 be the corresponding integration space. (In his work, Weba deals with more general lattice norms on integration spaces, but he only considers compact domains X and the topology of uniform convergence on X.) We investigate approximations of stochastic processes modeled by continuous linear operators T_n on $C\Psi(X, L^1)$. Continuity of the functions, of course, is meant with respect to the usual (symmetric) topology on L^1. As before, the operators should be positive, weakly E-commutative and stochastically simple (c.f. V.2.6). Since positivity is required, we endow L^1 as a locally convex cone with the neighborhoods $V = \{\rho B \mid \rho > 0\}$, where B denotes the positive unit ball of L^1; i.e. for $\eta, \phi \in L^1$ we have

$$\eta \le \phi + \rho B \quad \text{if and only if} \quad \|(\eta - \phi)^+\|_1 \le \rho.$$

(As usual, we set $\phi^+ = \phi \vee 0$ for $\phi \in L^1$, and f^+ for the function $x \to f(x)^+$.) This shows that for functions $f, g \in C\Psi(X, L^1)$, $\rho > 0$ and $\psi \in \Psi$ we have

$$f \le g + \rho B_\psi \quad \text{if and only if} \quad \psi(x)\,\|(f(x) - g(x))^+\|_1 \le \rho \quad \text{for all } x \in X.$$

Convergence of the operators is meant in this sense. As before, we identify $C\Psi(X)$ with a subspace of $C\Psi(X, L^1)$, and for $f \in C\Psi(X)$ and $\phi \in L^1$ we denote by $f \cdot \phi \in C\Psi(X, L^1)$ the function $x \to f(x)\phi$.

For the order of convergence of the operators T_n under consideration we assume:

$$T_n(p) \le p + O(\alpha_n), \quad pf_i \le T_n(pf_i) + O(\alpha_n), \quad T_n(pf_i^2) \le pf_i^2 + O(\alpha_n)$$

and

$$p \le T_n(p) + O(\delta_n).$$

Since L^1 is not a full cone we have to complement it. We set $Q = \{\phi + \rho B \mid \phi \in L^1, \rho \ge 0\}$. Q is ordered by inclusion and endowed with the neighborhood system $V = \{\rho B \mid \rho > 0\}$, whence a full locally convex cone. Its dual Q^* consists of all elements $\Phi \oplus r$ such that $r \ge 0$, and Φ is a positive function in L^∞ such that $\|\Phi\|_\infty \leq r$. Its operation on Q is defined by

$$\Phi \oplus r\,(\phi + \rho B) \;=\; \int_\Omega \phi\Phi\, d\mu + r\rho.$$

Now we have to extend our operators T_n which are defined on $C\Psi(X,L^1)$ to operators $\overline{T_n}$ on $C\Psi(X,Q)$:

Firstly, note that every function $F \in C\Psi(X,Q)$ may be expressed in a unique way by means of two functions $f \in C\Psi(X,L^1)$ and $e \in C\Psi(X)^+$ as

$$F(x) = f(x) + e(x)B \quad \text{for all } x \in X.$$

(To review this, recall the argument in the proof of Corollary 4.5.) As the operators T_n map $C(X)$ into itself and are positive, we would like to extend them by the formula

$$\overline{T_n}(F)(x) = T_n(f)(x) + T_n(e)(x)B \quad \text{for all } x \in X.$$

In order to justify this definition, i.e. the u-continuity of the operators $\overline{T_n}$ we proceed as follows:

Firstly, let $(A_i)_{i=1,\dots,n}$ be a disjoint partition of the measure space Ω into measurable subsets such that $\mu(A_i) = m_i$, i.e. $m_1 + \dots + m_n = 1$. By $\chi_i \in L^1$ denote the characteristic functions of the sets A_i. Let $e_i \in C\Psi(X)^+$ and consider the function

$$g = \sum_{i=1}^{n} e_i \cdot \chi_i \;\in C\Psi(X,L^1).$$

by our assumptions on the operators T_n this yields

$$T_n(g) = \sum_{i=1}^{n} T_n(e_i) \cdot \chi_i\,,$$

and for every $x \in X$ we have

$$\|g(x)\|_1 = \sum_{i=1}^{n} m_i e_i(x) \quad \text{and} \quad \|T_n(g)(x)\|_1 = \sum_{i=1}^{n} m_i T_n(e_i)(x)\,.$$

Thus for all functions $e \in C\Psi(X)^+$ and all functions g as above such that $g \leq e \cdot B$, this means

$$\sum_{i=1}^{n} m_i\, e_i(x) \leq e\,, \quad \text{whence} \quad \sum_{i=1}^{n} m_i\, T_n(e)(x) \leq T_n(e)\,.$$

and $T_n(g) \leq T_n(e) \cdot B$ as well. Now a standard argument using the definition of Nachbin cones, continuity and compactness (the functions ψf for $\psi \in \Psi$ and $f \in C\Psi(X,L^1)$ all vanish at infinity) yields that for every function $f \in C\Psi(X,L^1)$ such that $f \leq e \cdot B$, its positive part f^+ may be approximated by functions $g \leq e \cdot B$ of the above type. Summarizing, this yields that $f \leq e \cdot B$ for $f \in C\Psi(X,L^1)$ implies $T_n(f) \leq T_n(e) \cdot B$ in any case.

For a function $f \in C\Psi(X,L^1)$ now we denote by $E(f)$ the real-valued function $x \to \|(f(x))^+\|_1$ which is contained in $C\Psi(X)^+$. Then the preceding argument shows that

$$T_n(f) \leq T_n(E(f)) \cdot B$$

holds for all $f \in C\Psi(X,L^1)$.

Now secondly, let $\rho > 0$ and $\psi \in \Psi$. By u-continuity there are $\rho' > 0$ and $\phi \in \Psi$ such that for all $f, g \in C\Psi(X,L^1)$

$$f \leq g + \rho B_\psi \quad \text{implies} \quad T_n(f) \leq T_n(g) + \rho' B_\phi.$$

Let $F,G \in C\Psi(X,Q)$ such that $F \leq G+\rho B_\psi$; i.e. there are functions $f,g \in C\Psi(X,L^1)$ and $e,d \in C\Psi(X)^+$ such that $F = f+e\cdot B$ and $G = g+e\cdot B$ and

$$\psi(x)(f(x)+e(x)B) \leq \psi(x)(g(x)+d(x)B) + \rho B \qquad \text{for all } x \in X,$$

i.e.
$$\psi(x)\|(f-g)^+(x)\|_1 \leq \psi(x)(d-e)(x) + \rho B \qquad \text{for all } x \in X.$$

But the latter means $E(f-g) \leq (d-e)+\rho B_\psi$ and implies $T_n(E(f-g)) \leq T_n((d-e))+\rho B_\psi$.

i.e.
$$\psi(x)(T_n(E(f-g)(x)) \leq \psi(x)(T_n((d-e)(x))+\rho \qquad \text{for all } x \in X.$$

Finally, our first argument now yields

$$\psi(x)\,\|(T_n(f)-T_n(g))^+(x)\|_1 \leq \psi(x)(T_n(d)(x)-T_n(e)(x))+\rho' \qquad \text{for all } x \in X,$$

whence

$$\overline{T_n}(F) = T_n(f) + T_n(e)\cdot B \leq T_n(g) + T_n(d)\cdot B + \rho' B_\phi = \overline{T_n}(G) + \rho' B_\phi;$$

and the extended operators $\overline{T_n}$ are seen to be u-continuous, indeed.

Now Theorem 4.1 applies for the operators $\overline{T_n}$: We have $\overline{T_n}(E_B) \leq E_B + O(\alpha_n)$ for the set E_B of the functions

$$x \to p\,(f_i(x)-f_i(x_0))^2 B \quad \text{for all } f_i \in M,\ x_0 \in X.$$

We set $C = \{\phi+B \mid \phi \in L^1, \|\phi\| \leq 1\}$ and choose the subset $K = \{\Phi \oplus \|\phi\|_\infty \mid 0 \leq \phi \in L^\infty\}$ which is strictly separating for Q. For every function in $f \in C\Psi(X,Q)$ such that $\|f(x)\| \leq \lambda p(x)$ for some $\lambda > 0$ and all $x \in X$, the values of f/p are seen to be C^ω-smooth on K for $\omega(t) = t$. Finally, our assumptions guarantee $T_n(C) \leq C + O(\alpha_n)$ and $B \leq T_n(B) + O(\alpha_n)$.

Summarizing, this yields or final corollary:

4.10 Corollary. *Let (Ω, A, σ) be a probability space and L^1 the corresponding integration space. Let $(T_n)_{n \in N}$ be an equicontinuous sequence of positive linear weakly E-commutative and stochastically simple operators on $C\Psi(X,L^1)$. If*

$$T_n(p) \leq p+O(\alpha_n), \quad pf_i \leq T_n(pf_i)+O(\alpha_n), \quad T_n(pf_i^2) \leq pf_i^2+O(\alpha_n)$$

and $$p \leq T_n(p)+O(\delta_n) \quad \textit{for } i = 1,\ldots, N,$$

then $$T_n(f) \leq f+O(\max\{\omega(\alpha_n), \delta_n\})$$

for every function $f \in C\Psi(X, L^1)$ such that f/p is Δ^ω-continuous and $\|f(x)\|_1 \leq \lambda p(x)$ for some $\lambda > 0$ and all $x \in X$.

If the modulus function ω is of Lipschitz type, then an analogous statement holds for o-type order of convergence as well.

References

[1] **Alfsen, E.M.:** *Compact convex sets and boundary integrals.* Ergebnisse der Mathematik und ihrer Grenzgebiete 57, Springer Verlag, Heidelberg - Berlin - New York.

[2] **Alfsen, E.M. and Effros, E.G.:** *Structure in real Banach spaces.* Ann. Math. 96, I., 98-108 (1972).

[3] **Altomare, F.:** *On Korovkin type theorems in spaces of continuous complex-valued functions.* Bolletino U.M.I. (6) 1-B, 75-86 (1982).

[4] **Altomare, F.:** *On the Korovkin approximation theory in commutative Banach algebras.* Rendiconti Math. 2.4, 755-767 (1982).

[5] **Altomare, F.:** *On the universal convergence set.* Ann. Math. Pura Appl., IV. ser., 138, 223-243 (1984).

[6] **Altomare, F.:** *Nets of positive operators in spaces of continuous affine functions.* Boll. UMI, VII. ser., B.1, 217-233 (1987).

[7] **Altomare, F. and Campiti, M.:** *A bibliography on the Korovkin type approximation theory (1952-1987).* Instituto di Matematica, Università della Basilicata, Potenza, (Preprint).

[8] **Bauer, H.:** *Shilovscher Rand und Dirichletsches Problem.* Ann. Inst. Fourier 11, 89-136 (1961).

[9] **Bauer, H.:** *Theorems of Korovkin type for adapted spaces.* Ann. Inst. Fourier 23, fasc. 4, 245-260 (1973).

[10] **Bauer, H.:** *Funktionenkegel und Integralungleichungen.* Sitz. Ber. math. naturw. Kl. Bayer. Akad. Wiss. München 1977, 53-61 (1978).

[11] **Bauer, H. and Donner, K.:** *Korovkin Approximation in $C_0(X)$.* Math. Ann. 236, 225-237 (1978).

[12] **Behrens, H. and Lorentz, G.G.:** *Theorems of Korovkin type for positive linear operators on Banach lattices.* In: Approximation Theory (ed.:G.G. Lorentz), Academic Press, New York, 1-30 (1973).

[13] **Campiti, M.:** *Approximation of continuous set-valued functions in Fréchet spaces.* Univ. degli Studi, Bari, Preprint.

[14] **Campiti, M.:** *A Korovkin type theorem for set-valued Hausdorff continuous functions.* Le Matematiche, vol. XLII, fasc. I-II, 29-35 (1987).

[15] **Censor, E.:** *Quantitative results for positive linear approximation operators.* J. Approximation Theory 4, 442-450 (1971).

[16] **Choquet, G. and Deny, J.:** *Ensembles semi-réticulés et ensembles réticulés de fonctions continues.* J. Math. Pures Appl. 36, 179-189 (1957).

[17] **Cunningham jr., F.:** *L-structure in L-spaces.* Trans. Amer. Math. Soc. 95, 274-299 (1960).

[18] **De Vore, R.A.:** *The approximation of continuous functions by positive operators and Korovkin theorems.* Lecture Notes in Math. 293, Springer Verlag, Heidelberg - Berlin - New York.

[19] **Donner, K.:** *Extensions of positive operators and Korovkin Theorems.* Lecture Notes in Math. 904, Springer Verlag, Heidelberg - Berlin - New York.

[20] **Flösser, H.O., Irmisch, R. and Roth, W.:** *Infimum-stable convex cones and approximation.* Proc. London Math. Soc. (3) 42, 104-120 (1981).

[21] **Flösser, H.O. and Roth, W.:** *Korovkinhüllen in Funktionenräumen.* Math. Zeitschrift 166, 187-203 (1979).

[22] **Fuchssteiner, B. and Lusky, W.:** *Convex Cones.* North Holland Math. Studies 56, (1981).

[23] **Gadzhiev, A.D.:** *Theorems of Korovkin type.* Math. Notes 20, 995-998 (1976).

[24] **Gadzhiev, A.D.:** *Positive linear operators in weighted spaces of functions in several variables* (Russian). Izv. Akad. Nauk. Azerbaidzan SSSR, ser. Fiz.-Tehn. Mat. Nauk 1, no. 4, 32-37 (1980).

[26] **Hörmander, L.:** *Sur la fonction d'appui des ensembles convexes dans un espace localement convexe.* Arkiv Mat. 3, 181-186 (1954).

[27] **Hudzik, H., Musielak, J. and Urbansky, R.:** *Lattice properties of the space of convex closed and bounded subsets of a topological vector space.* Bull. Acad. Polon. Sci. Sér. Sci. Math 27, 157-162 (1979).

[28] **Keimel, K. and Roth, W.:** *A Korovkin type approximation theorem for set-valued functions.* Proc. Amer. Math. Soc. 104, 819-823 (1988).

[29] **Korovkin, P.P.:** *On convergence of linear positive operators in the space of continuous functions* (Russian). Dokl. Akad. Nauk SSSR 90, 961-964 (1953).

[30] **Korovkin, P.P.:** *Linear operators and approximation theory.* Russian Monographs and Texts on Advanced Mathematics and Physics, vol. III, Gordon and Breach Publ., New York (1960).

[31] **Lembcke, J.:** *Note zu "Funktionenkegel und Integralungleichungen" von H. Bauer.* Sitz. Ber. math.-naturw. Kl. Bayer. Akad. Wiss. München 1977, 139-142 (1978).

[32] **Michael, E.:** *Continuous selections I.* Ann. of Math. 63, 361-382 (1956).

[33] **Michael, E.:** *Convex structures and continuous selections.* Canadian J. Math. 11, 556-575 (1959).

[34] **Nachbin, L.:** *Topology and order.* Van Nostrand, Princeton (1965).

[35] **Nishishiraho, T.:** *Quantitative theorems on approximation processes of positive linear operators.* In: Multivariate Approximation Theory II (edited by W.Schempp and K. Zeller), ISNM 61, 297-311, Birkhäuser Verlag, Basel, Boston, Stuttgart.

[36] **Nishishiraho, T.:** *The degree of convergence of positive linear operators.* Tôhoku Math. J. 29, 81-89 (1977).

[37] **Nishishiraho, T.:** *Convergence of positive linear approximation processes.* Tôhoku Math. J. 35, 441-458 (1983).

[38] **Nishishiraho, T.:** *The convergence and saturation of iterations of positive linear operators.* Math. Z. 194, 397-404 (1987).

[39] **Pannenberg, M.:** *Korovkin approximation in Waelbroek-algebras.* Math. Ann. 274, 423-437 (1986).

[40] **Pannenberg, M.:** *Korovkin approximation in algebras of vector-valued functions.* Math. Inst. Univ. Münster, Preprint (1987).

[41] **Pannenberg, M.:** *Topics in qualitative Korovkin approximation.* Math. Inst. Univ. Münster, Preprint (1987).

[42] **Pinsker, A.G.:** *The space of convex sets of a locally convex space* (Russian). Trudy Leningrad. Inzh.-Ekon. In-ta 63, 13-17 (1966).

[43] **Prolla, J.B.:** *Approximation of vector-valued functions.* North Holland, Math. Studies 25 (1977).

[44] **Prolla, J.B.:** *Approximation of continuous convex-cone-valued functions by monotone operators.* Univ. Estual de Campinas Inst. Mat., Technical Report 27 (1990).

[45] **Prolla, J.B.:** *The Weierstrass-Stone theorem for convex-cone-valued functions.* Univ. Estual de Campinas Inst. Mat., Preprint.

[46] **Rabinovich, M.G.:** *An embedding theorem for spaces of convex sets* (Russian). Sibirsk. Mat. Zh. 8, 376-383 (1967). (English transl.) Siberian Math. J. 8, 275-279 (1967).

[47] **Rabinovich, M.G.:** *Some classes of spaces of convex sets and their extensions* (Russian). Sibirsk. Mat. Zh. 8, 1405-1415 (1967). (English transl.) Siberian Math. J. 8, 1064-1070 (1967).)

[48] **Rådström, H.:** *An embedding theorem for spaces of convex sets.* Proc. Amer. Math. Soc. 3, 165-169 (1952).

[49] **Roth, W.:** *Families of convex subsets and Korovkin type theorems in locally convex spaces.* Rev. Roumaine Math. Pures et Appl. 34, 329-346 (1989).

[50] **Shashkin, Yu. A. :** *On the convergence of linear positive operators in the space of continuous functions* (Russian). Dokl. Akad. Nauk SSSR 131, 525-527 (1960).

[51] **Shashkin, Yu. A. :** *Korovkin systems in spaces of continuous functions.* Am. Math. Soc. Transl. ser. 2, 54, 125-144 (1966).

[52] **Shashkin, Yu. A. :** *The Milman-Choquet boundary and approximation theory.* Functional Anal. Appl. 1, 170-171 (1967).

[53] **Scheffold, E.:** *Über Konvergenz linearer Operatoren.* Mathematica 20 (43), 193-198 (1975).

[54] **Schirotzek, W.:** *Extensions of additive non-negatively homogeneous functionals in quasilinear spaces.* Demonstratio Math. 10, 241-260 (1977).

[55] **Schmidt, K.D.:** *Embedding theorems for classes of convex sets.* Acta Appl. Math. 5, 209-237 (1986).

[56] **Singer, I.:** *Sur la meilleure approximation des fonctions abstraites continues à valeurs dans un espace de Banach.* Rev. Roum. Math. Pures et Appl. 2, 245-262 (1957).

[57] **Topsøe, F.:** *The naive approach to the Hahn-Banach Theorem.* Comment. Math. Special Issue 1, 315-330 (1978).

[58] **Urbanski, R.:** *A generalization of the Minkowsky-Rådström-Hörmander Theorem.* Bull. Acad. Polon. Sc., ser. sc. math., astr. et phys. 24, 709-725 (1976).

[59] **Vitale, R.A.:** *Approximation of convex set-valued functions.* J. Approximation Theory 26, 301-316 (1979).

[60] **Weba, M.:** *Korovkin systems of stochastic processes.* Math. Z. 192, 73-80 (1986).

[61] **Weba, M.:** *A quantitative Korovkin theorem for random functions with multivariate domain.* J. Approximation Theory 61, 74-87 (1990).

[62] **Wolff, M.:** *On the theory of approximation by positive linear operators in vector lattices.* In: Functional Analysis: Surveys and recent results (eds.: K. D. Bierstedt, B. Fuchssteiner), North Holland Math. Studies 27, 73-87 (1977).

[63] **Wulbert, D. E.:** *Convergence of operators and Korovkin's theorem.* J. Approximation Theory 1, 381-390 (1968).

Index

Lecture Notes in Mathematics

For information about Vols. 1–1312
please contact your bookseller or Springer-Verlag

Vol. 1313: F. Colonius. Optimal Periodic Control. VI, 177 pages. 1988.

Vol. 1314: A. Futaki. Kähler-Einstein Metrics and Integral Invariants. IV, 140 pages. 1988.

Vol. 1315: R.A. McCoy, I. Ntantu, Topological Properties of Spaces of Continuous, Functions. IV, 124 pages. 1988.

Vol. 1316: H. Korezlioglu, A.S. Ustunel (Eds.), Stochastic Analysis and Related Topics. Proceedings, 1986. V, 371 pages. 1988.

Vol. 1317: J. Lindenstrauss, V.D. Milman (Eds.), Geometric Aspects of Functional Analysis. Seminar, 1986–87. VII, 289 pages. 1988.

Vol. 1318: Y. Felix (Ed.), Algebraic Topology – Rational Homotopy. Proceedings, 1986. VIII, 245 pages. 1988.

Vol. 1319: M. Vuorinen, Conformal Geometry and Quasiregular Mappings. XIX, 209 pages. 1988.

Vol. 1320: H. Jürgensen, G. Lallement, H.J. Weinert (Eds.), Semigroups, Theory and Applications. Proceedings, 1986. X, 416 pages. 1988.

Vol. 1321: J. Azéma, P.A. Meyer, M. Yor (Eds.), Séminaire de Probabilités XXII. Proceedings. IV, 600 pages. 1988.

Vol. 1322: M. Métivier, S. Watanabe (Eds.), Stochastic Analysis. Proceedings, 1987. VII, 197 pages. 1988.

Vol. 1323: D.R. Anderson, H.J. Munkholm, Boundedly Controlled Topology. XII, 309 pages. 1988.

Vol. 1324: F. Cardoso, D.G. de Figueiredo, R. Iório, O. Lopes (Eds.), Partial Differential Equations. Proceedings, 1986. VIII, 433 pages. 1988.

Vol. 1325: A. Truman, I.M. Davies (Eds.), Stochastic Mechanics and Stochastic Processes. Proceedings, 1986. V, 220 pages. 1988.

Vol. 1326: P.S. Landweber (Ed.), Elliptic Curves and Modular Forms in Algebraic Topology. Proceedings, 1986. V, 224 pages. 1988.

Vol. 1327: W. Bruns, U. Vetter, Determinantal Rings. VII, 236 pages. 1988.

Vol. 1328: J.L. Bueso, P. Jara, B. Torrecillas (Eds.), Ring Theory. Proceedings, 1986. IX, 331 pages. 1988.

Vol. 1329: M. Alfaro, J.S. Dehesa, F.J. Marcellan, J.L. Rubio de Francia, J. Vinuesa (Eds.): Orthogonal Polynomials and their Applications. Proceedings, 1986. XV, 334 pages. 1988.

Vol. 1330: A. Ambrosetti, F. Gori, R. Lucchetti (Eds.), Mathematical Economics. Montecatini Terme 1986. Seminar. VII, 137 pages. 1988.

Vol. 1331: R. Bamón, R. Labarca, J. Palis Jr. (Eds.), Dynamical Systems, Valparaiso 1986. Proceedings. VI, 250 pages. 1988.

Vol. 1332: E. Odell, H. Rosenthal (Eds.), Functional Analysis. Proceedings. 1986–87. V, 202 pages. 1988.

Vol. 1333: A.S. Kechris, D.A. Martin, J.R. Steel (Eds.), Cabal Seminar 81–85. Proceedings, 1981–85. V, 224 pages. 1988.

Vol. 1334: Yu.G. Borisovich, Yu.E. Gliklikh (Eds.), Global Analysis – Studies and Applications III. V, 331 pages. 1988.

Vol. 1335: F. Guillén, V. Navarro Aznar, P. Pascual-Gainza, F. Puerta, Hyperrésolutions cubiques et descente cohomologique. XII, 192 pages. 1988.

Vol. 1336: B. Helffer, Semi-Classical Analysis for the Schrödinger Operator and Applications. V, 107 pages. 1988.

Vol. 1337: E. Sernesi (Ed.), Theory of Moduli. Seminar, 1985. VIII, 232 pages. 1988.

Vol. 1338: A.B. Mingarelli, S.G. Halvorsen. Non-Oscillation Domains of Differential Equations with Two Parameters. XI, 109 pages. 1988.

Vol. 1339: T. Sunada (Ed.), Geometry and Analysis of Manifolds. Proceedings, 1987. IX, 277 pages. 1988.

Vol. 1340: S. Hildebrandt, D.S. Kinderlehrer, M. Miranda (Eds.), Calculus of Variations and Partial Differential Equations. Proceedings, 1986. IX, 301 pages. 1988.

Vol. 1341: M. Dauge, Elliptic Boundary Value Problems on Corner Domains. VIII, 259 pages. 1988.

Vol. 1342: J.C. Alexander (Ed.), Dynamical Systems. Proceedings, 1986–87. VIII, 726 pages. 1988.

Vol. 1343: H. Ulrich, Fixed Point Theory of Parametrized Equivariant Maps. VII, 147 pages. 1988.

Vol. 1344: J. Král, J. Lukes, J. Netuka, J. Vesely´ (Eds.), Potential Theory – Surveys and Problems. Proceedings, 1987. VIII, 271 pages. 1988.

Vol. 1345: X. Gomez-Mont, J. Seade, A. Verjovski (Eds.), Holomorphic Dynamics. Proceedings, 1986. VII. 321 pages. 1988.

Vol. 1346: O.Ya. Viro (Ed.), Topology and Geometry – Rohlin Seminar. XI, 581 pages. 1988.

Vol. 1347: C. Preston, Iterates of Piecewise Monotone Mappings on an Interval. V, 166 pages. 1988.

Vol. 1348: F. Borceux (Ed.), Categorical Algebra and its Applications. Proceedings, 1987. VIII, 375 pages. 1988.

Vol. 1349: E. Novak, Deterministic and Stochastic Error Bounds in Numerical Analysis. V, 113 pages. 1988.

Vol. 1350: U. Koschorke (Ed.), Differential Topology Proceedings, 1987, VI, 269 pages. 1988.

Vol. 1351: I. Laine, S. Rickman, T. Sorvali (Eds.), Complex Analysis, Joensuu 1987. Proceedings. XV, 378 pages. 1988.

Vol. 1352: L.L. Avramov, K.B. Tchakerian (Eds.), Algebra – Some Current Trends. Proceedings. 1986. IX, 240 Seiten. 1988.

Vol. 1353: R.S. Palais, Ch.-l. Teng, Critical Point Theory and Submanifold Geometry. X, 272 pages. 1988.

Vol. 1354: A. Gómez, F. Guerra, M.A. Jiménez, G. López (Eds.), Approximation and Optimization. Proceedings, 1987. VI, 280 pages. 1988.

Vol. 1355: J. Bokowski, B. Sturmfels, Computational Synthetic Geometry. V, 168 pages. 1989.

Vol. 1356: H. Volkmer, Multiparameter Eigenvalue Problems and Expansion Theorems. VI, 157 pages. 1988.

Vol. 1357: S. Hildebrandt, R. Leis (Eds.), Partial Differential Equations and Calculus of Variations. VI, 423 pages. 1988.

Vol. 1358: D. Mumford, The Red Book of Varieties and Schemes. V, 309 pages. 1988.

Vol. 1359: P. Eymard, J.-P. Pier (Eds.) Harmonic Analysis. Proceedings, 1987. VIII, 287 pages. 1988.

Vol. 1360: G. Anderson, C. Greengard (Eds.), Vortex Methods. Proceedings, 1987. V, 141 pages. 1988.

Vol. 1361: T. tom Dieck (Ed.), Algebraic Topology and Transformation Groups. Proceedings. 1987. VI, 298 pages. 1988.

Vol. 1362: P. Diaconis, D. Elworthy, H. Föllmer, E. Nelson, G.C. Papanicolaou, S.R.S. Varadhan. École d´ Été de Probabilités de Saint-Flour XV–XVII. 1985–87 Editor: P.L. Hennequin. V, 459 pages. 1988.

Vol. 1363: P.G. Casazza, T.J. Shura, Tsirelson´s Space. VIII, 204 pages. 1988.

Vol. 1364: R.R. Phelps, Convex Functions, Monotone Operators and Differentiability. IX, 115 pages. 1989.

Vol. 1365: M. Giaquinta (Ed.), Topics in Calculus of Variations. Seminar, 1987. X, 196 pages. 1989.

Vol. 1366: N. Levitt, Grassmannians and Gauss Maps in PL-Topology. V, 203 pages. 1989.

Vol. 1367: M. Knebusch, Weakly Semialgebraic Spaces. XX, 376 pages. 1989.

Vol. 1368: R. Hübl, Traces of Differential Forms and Hochschild Homology. III, 111 pages. 1989.

Vol. 1369: B. Jiang, Ch.-K. Peng, Z. Hou (Eds.), Differential Geometry and Topology. Proceedings, 1986–87. VI, 366 pages. 1989.

Vol. 1370: G. Carlsson, R.L. Cohen, H.R. Miller, D.C. Ravenel (Eds.), Algebraic Topology. Proceedings, 1986. IX, 456 pages. 1989.

Vol. 1371: S. Glaz, Commutative Coherent Rings. XI, 347 pages. 1989.

Vol. 1372: J. Azéma, P.A. Meyer, M. Yor (Eds.), Séminaire de Probabilités XXIII. Proceedings. IV, 583 pages. 1989.

Vol. 1373: G. Benkart, J.M. Osborn (Eds.), Lie Algebras. Madison 1987. Proceedings. V, 145 pages. 1989.

Vol. 1374: R.C. Kirby, The Topology of 4-Manifolds. VI, 108 pages. 1989.

Vol. 1375: K. Kawakubo (Ed.), Transformation Groups. Proceedings, 1987. VIII, 394 pages, 1989.

Vol. 1376: J. Lindenstrauss, V.D. Milman (Eds.), Geometric Aspects of Functional Analysis. Seminar (GAFA) 1987–88. VII, 288 pages. 1989.

Vol. 1377: J.F. Pierce, Singularity Theory, Rod Theory, and Symmetry-Breaking Loads. IV, 177 pages. 1989.

Vol. 1378: R.S. Rumely, Capacity Theory on Algebraic Curves. III, 437 pages. 1989.

Vol. 1379: H. Heyer (Ed.), Probability Measures on Groups IX. Proceedings, 1988. VIII, 437 pages. 1989.

Vol. 1380: H.P. Schlickewei, E. Wirsing (Eds.), Number Theory, Ulm 1987. Proceedings. V, 266 pages. 1989.

Vol. 1381: J.-O. Strömberg, A. Torchinsky, Weighted Hardy Spaces. V, 193 pages. 1989.

Vol. 1382: H. Reiter, Metaplectic Groups and Segal Algebras. XI, 128 pages. 1989.

Vol. 1383: D.V. Chudnovsky, G.V. Chudnovsky, H. Cohn, M.B. Nathanson (Eds.), Number Theory, New York 1985–88. Seminar. V, 256 pages. 1989.

Vol. 1384: J. Garcia-Cuerva (Ed.), Harmonic Analysis and Partial Differential Equations. Proceedings, 1987. VII, 213 pages. 1989.

Vol. 1385: A.M. Anile, Y. Choquet-Bruhat (Eds.), Relativistic Fluid Dynamics. Seminar, 1987. V, 308 pages. 1989.

Vol. 1386: A. Bellen, C.W. Gear, E. Russo (Eds.), Numerical Methods for Ordinary Differential Equations. Proceedings, 1987. VII, 136 pages. 1989.

Vol. 1387: M. Petkovi´c, Iterative Methods for Simultaneous Inclusion of Polynomial Zeros. X, 263 pages. 1989.

Vol. 1388: J. Shinoda, T.A. Slaman, T. Tugué (Eds.), Mathematical Logic and Applications. Proceedings, 1987. V, 223 pages. 1989.

Vol. 1000: Second Edition. H. Hopf, Differential Geometry in the Large. VII, 184 pages. 1989.

Vol. 1389: E. Ballico, C. Ciliberto (Eds.), Algebraic Curves and Projective Geometry. Proceedings, 1988. V, 288 pages. 1989.

Vol. 1390: G. Da Prato, L. Tubaro (Eds.), Stochastic Partial Differential Equations and Applications II. Proceedings, 1988. VI, 258 pages. 1989.

Vol. 1391: S. Cambanis, A. Weron (Eds.), Probability Theory on Vector Spaces IV. Proceedings, 1987. VIII, 424 pages. 1989.

Vol. 1392: R. Silhol, Real Algebraic Surfaces. X, 215 pages. 1989.

Vol. 1393: N. Bouleau, D. Feyel, F. Hirsch, G. Mokobodzki (Eds.), Séminaire de Théorie du Potentiel Paris, No. 9. Proceedings. VI, 265 pages. 1989.

Vol. 1394: T.L. Gill, W.W. Zachary (Eds.), Nonlinear Semigroups, Partial Differential Equations and Attractors. Proceedings, 1987. IX, 233 pages. 1989.

Vol. 1395: K. Alladi (Ed.), Number Theory, Madras 1987. Proceedings. VII, 234 pages. 1989.

Vol. 1396: L. Accardi, W. von Waldenfels (Eds.), Quantum Probability and Applications IV. Proceedings, 1987. VI, 355 pages. 1989.

Vol. 1397: P.R. Turner (Ed.), Numerical Analysis and Parallel Processing. Seminar, 1987. VI, 264 pages. 1989.

Vol. 1398: A.C. Kim, B.H. Neumann (Eds.), Groups – Korea 1988. Proceedings. V, 189 pages. 1989.

Vol. 1399: W.-P. Barth, H. Lange (Eds.), Arithmetic of Complex Manifolds. Proceedings, 1988. V, 171 pages. 1989.

Vol. 1400: U. Jannsen. Mixed Motives and Algebraic K-Theory. XIII, 246 pages. 1990.

Vol. 1401: J. Steprans, S. Watson (Eds.), Set Theory and its Applications. Proceedings, 1987. V, 227 pages. 1989.

Vol. 1402: C. Carasso, P. Charrier, B. Hanouzet, J.-L. Joly (Eds.), Nonlinear Hyperbolic Problems. Proceedings, 1988. V, 249 pages. 1989.

Vol. 1403: B. Simeone (Ed.), Combinatorial Optimization. Seminar, 1986. V, 314 pages. 1989.

Vol. 1404: M.-P. Malliavin (Ed.), Séminaire d´Algèbre Paul Dubreil et Marie-Paul Malliavin. Proceedings, 1987–1988. IV, 410 pages. 1989.

Vol. 1405: S. Dolecki (Ed.), Optimization. Proceedings, 1988. V, 223 pages. 1989. Vol. 1406: L. Jacobsen (Ed.), Analytic Theory of Continued Fractions III. Proceedings, 1988. VI, 142 pages. 1989.

Vol. 1407: W. Pohlers, Proof Theory. VI, 213 pages. 1989.

Vol. 1408: W. Lück, Transformation Groups and Algebraic K-Theory. XII, 443 pages. 1989.

Vol. 1409: E. Hairer, Ch. Lubich, M. Roche. The Numerical Solution of Differential-Algebraic Systems by Runge-Kutta Methods. VII, 139 pages. 1989.

Vol. 1410: F.J. Carreras, O. Gil-Medrano, A.M. Naveira (Eds.), Differential Geometry. Proceedings, 1988. V, 308 pages. 1989.

Vol. 1411: B. Jiang (Ed.), Topological Fixed Point Theory and Applications. Proceedings. 1988. VI, 203 pages. 1989.

Vol. 1412: V.V. Kalashnikov, V.M. Zolotarev (Eds.), Stability Problems for Stochastic Models. Proceedings, 1987. X, 380 pages. 1989.

Vol. 1413: S. Wright, Uniqueness of the Injective III$_1$ Factor. III, 108 pages. 1989.

Vol. 1414: E. Ramirez de Arellano (Ed.), Algebraic Geometry and Complex Analysis. Proceedings, 1987. VI, 180 pages. 1989.

Vol. 1415: M. Langevin, M. Waldschmidt (Eds.), Cinquante Ans de Polynômes. Fifty Years of Polynomials. Proceedings, 1988. IX, 235 pages. 1990.

Vol. 1416: C. Albert (Ed.), Géométrie Symplectique et Mécanique. Proceedings, 1988. V, 289 pages. 1990.

Vol. 1417: A.J. Sommese, A. Biancofiore, E.L. Livorni (Eds.), Algebraic Geometry. Proceedings, 1988. V, 320 pages. 1990.

Vol. 1418: M. Mimura (Ed.), Homotopy Theory and Related Topics. Proceedings, 1988. V, 241 pages. 1990.

Vol. 1419: P.S. Bullen, P.Y. Lee, J.L. Mawhin, P. Muldowney, W.F. Pfeffer (Eds.), New Integrals. Proceedings, 1988. V, 202 pages. 1990.

Vol. 1420: M. Galbiati, A. Tognoli (Eds.), Real Analytic Geometry. Proceedings, 1988. IV, 366 pages. 1990.

Vol. 1421: H.A. Biagioni, A Nonlinear Theory of Generalized Functions, XII, 214 pages. 1990.

Vol. 1422: V. Villani (Ed.), Complex Geometry and Analysis. Proceedings, 1988. V, 109 pages. 1990.

Vol. 1423: S.O. Kochman, Stable Homotopy Groups of Spheres: A Computer-Assisted Approach. VIII, 330 pages. 1990.

Vol. 1424: F.E. Burstall, J.H. Rawnsley, Twistor Theory for Riemannian Symmetric Spaces. III, 112 pages. 1990.

Vol. 1425: R.A. Piccinini (Ed.), Groups of Self-Equivalences and Related Topics. Proceedings, 1988. V, 214 pages. 1990.

Vol. 1426: J. Azéma, P.A. Meyer, M. Yor (Eds.), Séminaire de Probabilités XXIV, 1988/89. V, 490 pages. 1990.

Vol. 1427: A. Ancona, D. Geman, N. Ikeda, École d'Eté de Probabilités de Saint Flour XVIII, 1988. Ed.: P.L. Hennequin. VII, 330 pages. 1990.

Vol. 1428: K. Erdmann, Blocks of Tame Representation Type and Related Algebras. XV. 312 pages. 1990.

Vol. 1429: S. Homer, A. Nerode, R.A. Platek, G.E. Sacks, A. Scedrov, Logic and Computer Science. Seminar, 1988. Editor: P. Odifreddi. V, 162 pages. 1990.

Vol. 1430: W. Bruns, A. Simis (Eds.), Commutative Algebra. Proceedings. 1988. V, 160 pages. 1990.

Vol. 1431: J.G. Heywood, K. Masuda, R. Rautmann, V.A. Solonnikov (Eds.), The Navier-Stokes Equations – Theory and Numerical Methods. Proceedings, 1988. VII, 238 pages. 1990.

Vol. 1432: K. Ambos-Spies, G.H. Müller, G.E. Sacks (Eds.), Recursion Theory Week. Proceedings, 1989. VI, 393 pages. 1990.

Vol. 1433: S. Lang, W. Cherry, Topics in Nevanlinna Theory. II, 174 pages. 1990.

Vol. 1434: K. Nagasaka, E. Fouvry (Eds.), Analytic Number Theory. Proceedings, 1988. VI, 218 pages. 1990.

Vol. 1435: St. Ruscheweyh, E.B. Saff, L.C. Salinas, R.S. Varga (Eds.), Computational Methods and Function Theory. Proceedings, 1989. VI, 211 pages. 1990.

Vol. 1436: S. Xambó-Descamps (Ed.), Enumerative Geometry. Proceedings, 1987. V, 303 pages. 1990.

Vol. 1437: H. Inassaridze (Ed.), K-theory and Homological Algebra. Seminar, 1987–88. V, 313 pages. 1990.

Vol. 1438: P.G. Lemarié (Ed.) Les Ondelettes en 1989. Seminar. IV, 212 pages. 1990.

Vol. 1439: E. Bujalance, J.J. Etayo, J.M. Gamboa, G. Gromadzki. Automorphism Groups of Compact Bordered Klein Surfaces: A Combinatorial Approach. XIII, 201 pages. 1990.

Vol. 1440: P. Latiolais (Ed.), Topology and Combinatorial Groups Theory. Seminar, 1985–1988. VI, 207 pages. 1990.

Vol. 1441: M. Coornaert, T. Delzant, A. Papadopoulos. Géométrie et théorie des groupes. X, 165 pages. 1990.

Vol. 1442: L. Accardi, M. von Waldenfels (Eds.), Quantum Probability and Applications V. Proceedings, 1988. VI, 413 pages. 1990.

Vol. 1443: K.H. Dovermann, R. Schultz, Equivariant Surgery Theories and Their Periodicity Properties. VI, 227 pages. 1990.

Vol. 1444: H. Korezlioglu, A.S. Ustunel (Eds.), Stochastic Analysis and Related Topics VI. Proceedings, 1988. V, 268 pages. 1990.

Vol. 1445: F. Schulz, Regularity Theory for Quasilinear Elliptic Systems and – Monge Ampère Equations in Two Dimensions. XV, 123 pages. 1990.

Vol. 1446: Methods of Nonconvex Analysis. Seminar, 1989. Editor: A. Cellina. V, 206 pages. 1990.

Vol. 1447: J.-G. Labesse, J. Schwermer (Eds), Cohomology of Arithmetic Groups and Automorphic Forms. Proceedings, 1989. V, 358 pages. 1990.

Vol. 1448: S.K. Jain, S.R. López-Permouth (Eds.), Non-Commutative Ring Theory. Proceedings. 1989. V, 166 pages. 1990.

Vol. 1449: W. Odyniec, G. Lewicki, Minimal Projections in Banach Spaces. VIII, 168 pages. 1990.

Vol. 1450: H. Fujita, T. Ikebe, S.T. Kuroda (Eds.), Functional-Analytic Methods for Partial Differential Equations. Proceedings, 1989. VII, 252 pages. 1990.

Vol. 1451: L. Alvarez-Gaumé, E. Arbarello, C. De Concini, N.J. Hitchin, Global Geometry and Mathematical Physics. Montecatini Terme 1988. Seminar. Editors: M. Francaviglia, F. Gherardelli. IX, 197 pages. 1990.

Vol. 1452: E. Hlawka, R.F. Tichy (Eds.), Number-Theoretic Analysis. Seminar, 1988–89. V, 220 pages. 1990.

Vol. 1453: Yu.G. Borisovich, Yu.E. Gliklikh (Eds.), Global Analysis – Studies and Applications IV. V, 320 pages. 1990.

Vol. 1454: F. Baldassari, S. Bosch, B. Dwork (Eds.), p-adic Analysis. Proceedings, 1989. V, 382 pages. 1990.

Vol. 1455: J.-P. Françoise, R. Roussarie (Eds.), Bifurcations of Planar Vector Fields. Proceedings, 1989. VI, 396 pages. 1990.

Vol. 1456: L.G. Kovács (Ed.), Groups – Canberra 1989. Proceedings. XII, 198 pages. 1990.

Vol. 1457: O. Axelsson, L.Yu. Kolotilina (Eds.), Preconditioned Conjugate Gradient Methods. Proceedings, 1989. V, 196 pages. 1990.

Vol. 1458: R. Schaaf, Global Solution Branches of Two Point Boundary Value Problems. XIX, 141 pages. 1990.

Vol. 1459: D. Tiba, Optimal Control of Nonsmooth Distributed Parameter Systems. VII, 159 pages. 1990.

Vol. 1460: G. Toscani, V. Boffi, S. Rionero (Eds.), Mathematical Aspects of Fluid Plasma Dynamics. Proceedings, 1988. V, 221 pages. 1991.

Vol. 1461: R. Gorenflo, S. Vessella, Abel Integral Equations. VII, 215 pages. 1991.

Vol. 1462: D. Mond, J. Montaldi (Eds.), Singularity Theory and its Applications. Warwick 1989, Part I. VIII, 405 pages. 1991.

Vol. 1463: R. Roberts, I. Stewart (Eds.), Singularity Theory and its Applications. Warwick 1989, Part II. VIII, 322 pages. 1991.

Vol. 1464: D. L. Burkholder, E. Pardoux, A. Sznitman, Ecole d'Eté de Probabilités de Saint- Flour XIX-1989. Editor: P. L. Hennequin. VI, 256 pages. 1991.

Vol. 1465: G. David, Wavelets and Singular Integrals on Curves and Surfaces. X, 107 pages. 1991.

Vol. 1466: W. Banaszczyk, Additive Subgroups of Topological Vector Spaces. VII, 178 pages. 1991.

Vol. 1467: W. M. Schmidt, Diophantine Approximations and Diophantine Equations. VIII, 217 pages. 1991.

Vol. 1468: J. Noguchi, T. Ohsawa (Eds.), Prospects in Complex Geometry. Proceedings, 1989. VII, 421 pages. 1991.

Vol. 1469: J. Lindenstrauss, V. D. Milman (Eds.), Geometric Aspects of Functional Analysis. Seminar 1989-90. XI, 191 pages. 1991.

Vol. 1470: E. Odell, H. Rosenthal (Eds.), Functional Analysis. Proceedings, 1987-89. VII, 199 pages. 1991.

Vol. 1471: A. A. Panchishkin, Non-Archimedean L-Functions of Siegel and Hilbert Modular Forms. VII, 157 pages. 1991.

Vol. 1472: T. T. Nielsen, Bose Algebras: The Complex and Real Wave Representations. V, 132 pages. 1991.

Vol. 1473: Y. Hino, S. Murakami, T. Naito, Functional Differential Equations with Infinite Delay. X, 317 pages. 1991.

Vol. 1474: S. Jackowski, B. Oliver, K. Pawałowski (Eds.), Algebraic Topology, Poznań 1989. Proceedings. VIII, 397 pages. 1991.

Vol. 1475: S. Busenberg, M. Martelli (Eds.), Delay Differential Equations and Dynamical Systems. Proceedings, 1990. VIII, 249 pages. 1991.

Vol. 1476: M. Bekkali, Topics in Set Theory. VII, 120 pages. 1991.

Vol. 1477: R. Jajte, Strong Limit Theorems in Noncommutative L_2-Spaces. X, 113 pages. 1991.

Vol. 1478: M.-P. Malliavin (Ed.), Topics in Invariant Theory. Seminar 1989-1990. VI, 272 pages. 1991.

Vol. 1479: S. Bloch, I. Dolgachev, W. Fulton (Eds.), Algebraic Geometry. Proceedings, 1989. VII, 300 pages. 1991.

Vol. 1480: F. Dumortier, R. Roussarie, J. Sotomayor, H. Żoładek, Bifurcations of Planar Vector Fields: Nilpotent Singularities and Abelian Integrals. VIII, 226 pages. 1991.

Vol. 1481: D. Ferus, U. Pinkall, U. Simon, B. Wegner (Eds.), Global Differential Geometry and Global Analysis. Proceedings, 1991. VIII, 283 pages. 1991.

Vol. 1482: J. Chabrowski, The Dirichlet Problem with L^2-Boundary Data for Elliptic Linear Equations. VI, 173 pages. 1991.

Vol. 1483: E. Reithmeier, Periodic Solutions of Nonlinear Dynamical Systems. VI, 171 pages. 1991.

Vol. 1484: H. Delfs, Homology of Locally Semialgebraic Spaces. IX, 136 pages. 1991.

Vol. 1485: J. Azéma, P. A. Meyer, M. Yor (Eds.), Séminaire de Probabilités XXV. VIII, 440 pages. 1991.

Vol. 1486: L. Arnold, H. Crauel, J.-P. Eckmann (Eds.), Lyapunov Exponents. Proceedings, 1990. VIII, 365 pages. 1991.

Vol. 1487: E. Freitag, Singular Modular Forms and Theta Relations. VI, 172 pages. 1991.

Vol. 1488: A. Carboni, M. C. Pedicchio, G. Rosolini (Eds.), Category Theory. Proceedings, 1990. VII, 494 pages. 1991.

Vol. 1489: A. Mielke, Hamiltonian and Lagrangian Flows on Center Manifolds. X, 140 pages. 1991.

Vol. 1490: K. Metsch, Linear Spaces with Few Lines. XIII, 196 pages. 1991.

Vol. 1491: E. Lluis-Puebla, J.-L. Loday, H. Gillet, C. Soulé, V. Snaith, Higher Algebraic K-Theory: an overview. IX, 164 pages. 1992.

Vol. 1492: K. R. Wicks, Fractals and Hyperspaces. VIII, 168 pages. 1991.

Vol. 1493: E. Benoît (Ed.), Dynamic Bifurcations. Proceedings, Luminy 1990. VII, 219 pages. 1991.

Vol. 1494: M.-T. Cheng, X.-W. Zhou, D.-G. Deng (Eds.), Harmonic Analysis. Proceedings, 1988. IX, 226 pages. 1991.

Vol. 1495: J. M. Bony, G. Grubb, L. Hörmander, H. Komatsu, J. Sjöstrand, Microlocal Analysis and Applications. Montecatini Terme, 1989. Editors: L. Cattabriga, L. Rodino. VII, 349 pages. 1991.

Vol. 1496: C. Foias, B. Francis, J. W. Helton, H. Kwakernaak, J. B. Pearson, H_∞-Control Theory. Como, 1990. Editors: E. Mosca, L. Pandolfi. VII, 336 pages. 1991.

Vol. 1497: G. T. Herman, A. K. Louis, F. Natterer (Eds.), Mathematical Methods in Tomography. Proceedings 1990. X, 268 pages. 1991.

Vol. 1498: R. Lang, Spectral Theory of Random Schrödinger Operators. X, 125 pages. 1991.

Vol. 1499: K. Taira, Boundary Value Problems and Markov Processes. IX, 132 pages. 1991.

Vol. 1500: J. P. Serre, Lie Algebras and Lie Groups. VII, 168 pages. 1992.

Vol. 1501: A. De Masi, E. Presutti, Mathematical Methods for Hydrodynamic Limits. IX, 196 pages. 1991.

Vol. 1502: C. Simpson, Asymptotic Behavior of Monodromy. V, 139 pages. 1991.

Vol. 1503: S. Shokranian, The Selberg-Arthur Trace Formula (Lectures by J. Arthur). VII, 97 pages. 1991.

Vol. 1504: J. Cheeger, M. Gromov, C. Okonek, P. Pansu, Geometric Topology: Recent Developments. Editors: P. de Bartolomeis, F. Tricerri. VII, 197 pages. 1991.

Vol. 1505: K. Kajitani, T. Nishitani, The Hyperbolic Cauchy Problem. VII, 168 pages. 1991.

Vol. 1506: A. Buium, Differential Algebraic Groups of Finite Dimension. XV, 145 pages. 1992.

Vol. 1507: K. Hulek, T. Peternell, M. Schneider, F.-O. Schreyer (Eds.), Complex Algebraic Varieties. Proceedings, 1990. VII, 179 pages. 1992.

Vol. 1508: M. Vuorinen (Ed.), Quasiconformal Space Mappings. A Collection of Surveys 1960-1990. IX, 148 pages. 1992.

Vol. 1509: J. Aguadé, M. Castellet, F. R. Cohen (Eds.), Algebraic Topology - Homotopy and Group Cohomology. Proceedings, 1990. X, 330 pages. 1992.

Vol. 1510: P. P. Kulish (Ed.), Quantum Groups. Proceedings, 1990. XII, 398 pages. 1992.

Vol. 1511: D. Singh, B. S. Yadav (Eds.), Functional Analysis and Operator Theory. Proceedings, 1990. VIII, 223 pages. 1992.

Vol. 1512: L. M. Adleman, M.-D. A. Huang, Primality Testing and Abelian Varieties Over Finite Fields. VII, 142 pages. 1992.

Vol. 1513: L. S. Block, W. A. Coppel, Dynamics in One Dimension. VIII, 249 pages. 1992.

Vol. 1514: U. Krengel, K. Richter, V. Warstat (Eds.), Ergodic Theory and Related Topics III, Proceedings, 1990. VIII, 236 pages. 1992.

Vol. 1515: E. Ballico, F. Catanese, C. Ciliberto (Eds.), Classification of Irregular Varieties. Proceedings, 1990. VII, 149 pages. 1992.

Vol. 1517: K. Keimel, W. Roth, Ordered Cones and Approximation. VI, 134 pages. 1992.